Environment and Technology

Environment and Technology

Vierah Hulley

Delve
PUBLISHING

www.delvepublishing.com

Environment and Technology

Vierah Hulley

Delve Publishing
224 Shoreacres Road
Burlington, ON L7L 2H2
Canada
www.delvepublishing.com

Email: orders@arclereducation.com

ISBN: 978-1-77407-754-2 (Hardcover)

ABOUT THE AUTHOR

Vierah Hulley obtained her Ph.D. degree in Geohydrology from the University of the Free State, South Africa, in 2013. Her expertise is in Environmental and Earth Sciences and she has an extensive vocational background in the areas of natural resources management, environmental geochemistry, contaminated site remediation, sustainable development, and spatial analytics. She is the founder and Managing Director of HL Nexus: The Data Imaginarium; a company dedicated to optimizing natural resources management with data science.

TABLE OF CONTENTS

LIST OF FIGURES

LIST OF TABLES

LIST OF ABBREVIATIONS

AHAM	Association of Home Appliance Manufacturers
AI	artificial intelligence
CADR	clean air delivery rate
CSP	concentrated solar thermal plants
DOP	dioctyl phthalate
EPA	Environmental Protection Agency
GIGO	garbage in garbage out
GIS	geographic information systems
HEPA	high-efficiency particulate air
HTTD	high temperature thermal desorption
IoT	internet of things
IT	information technology
LTTD	low temperature thermal desorption
NAAQS	National Ambient Air Quality Standards
geoAI	geospatial artificial intelligence
PV	photovoltaic
SUVs	sport utility vehicles
ULPA	ultra-low penetration air
UV	ultraviolet
VAM	vesicular arbuscular mycorrhizal

PREFACE

Environmental management technologies constitute a myriad of tools and approaches used by various stakeholders in our society to mitigate the destruction of our environment. These technologies are also known as preventive environmental management (PEM). Human activities have contributed a large percentage to the degradation of our environment and therefore a way to counter this had to be found. Advancement in technology has helped in this quest to remediate the environment.

Technology has both negative and positive effects on the environment and therefore other structural measures should be put in place if we are to achieve a sustainable environment for the future generations. These technologies can be applied in manufacturing industries to produce cleaner products, add value to waste, minimize waste and reduce recalcitrance of these wastes. The appropriate application of the above-mentioned technologies can lead to safer workplaces, healthier communities, and a sustainable environment.

The global and domestic demand for an efficient society has been a very valuable driver for adopting new technologies and improved practices of production. This scenario provides a very opportune moment for industries to adopt these new technologies. Small and medium-sized companies have been playing a wild card on environmental management but with the adoption of new technologies the case will be soon resolved. The market, governments, and civil society organizations have been pressuring companies to adopt environmentally friendly methods of production.

Companies have focused solely on the embracing of technologies 'end of pipe' and then integrated technologies in the manufacturing process. This approach has seen the dwindling of companies resources due to lack of return of the invested capital. The saving grace for adopting technology in environmental management is that its long-term benefits are immense. With time the knowledge gain will be useful in minimizing waste and reducing environmental pollution. This then will help the companies reduce complex and expensive technologies applied on the treatment of the final product and the production costs by being more efficient.

The spheres of applying technology to manage the environment are many. For example, instead of relying on fossil fuel to generate energy why not embrace wind and solar power. These options are safer, cleaner, and cheaper. Low waste technologies increase the efficiency of raw materials, energy, and water through minimizing the generation of waste products or recycling them.

The volume has 10 chapters, and each chapter discusses about a specific topic under environment and technology. Chapter 1 discusses about introduction to environment and technology; Chapter 2 is about renewable energy technologies and the environment; Chapter 3 discusses about water purification technologies; Chapter 4 discusses about air purification technologies and the environment; Chapter 5 presents information about environmental remediation technologies; Chapter 6 discusses about eGAIN forecasting; Chapter 7 offers details about soil conservation technologies; Chapter 8 offer information on flood prevention technologies; Chapter 9 discusses about environmental remote sensing; and lastly, Chapter 10 is all about drones and environment.

Scientists and engineers are now looking for new ways for addressing issues regarding resources conservation and waste management. In this regard they could play a pivotal role on transition to cleaner production and reduction of waste production. Another promising technology that can safely reduce over-reliance on fossil fuels is thermo-depolymerization. This is a process that can turn any carbon-based waste, ranging from used tires to turkey guts, to produce petroleum. Experts working on this technology claim that a ton of turkey waste can be converted into 600 pounds of petroleum.

In conclusion, therefore, environmental management should pay closer attention to such technological advancement.

CHAPTER 1

Introduction to Environment and Technology

CONTENTS

1.1. INTRODUCTION

The Industrial Revolution spurred amazing gains in financial success. From 1870 to 1910, per capita revenue in the U.S. rose by nearly 40%, while the value of production output rose sevenfold. Still, rapid industrialization that was left following its wake dimmed noontime skies, loud, and unsafe machinery, plus severely affected living conditions (Figure 1.1).

Figure 1.1: Eco-friendly technologies ensure a healthy environment for everyone.

Source: http://www.allwritersdestination.com/eco-friendly-technologies-makes-our-environment-healthy-dwight-smith-on-tray-media/.

On average, it took almost three successive generations before the initial concerted efforts were done to bring contamination under control, though once started, development has been forthcoming. The air quality index for the U.S now reveals a 42% increment since 1980. The number of U.S. cities that fail to meet countrywide air quality criteria for ozone reservations has further been dropping, from 199 cities in 1990 to 70 cities by 1995. Vehicle emissions of 6 major air pollutants including sulfur dioxide, nitrogen oxides, carbon monoxide, greenhouse gases, particulate matter, and volatile organic compounds have reduced 31%, even while the amount of car miles driven has doubled (Yeoshua, 2005).

Having played a role in the pollution of the Earth, it's land, atmosphere, and water systems for well over a century, technology is currently showing potential in environmental proactively mitigating the effects of pollution and remediating the harmful effects after pollution has occurred. Technological innovations primarily aimed at decreasing pollution levels-from cleaner manufacturing systems to flue gas scrappers to catalytic converters-currently figure highly in mitigating a number of the development pains of a progressively technological world.

Technology, basically, is a double-edged sword that's capable of contributing to and undoing damage for environmental quality. Here, we inspect technology and environment based on four key aspects which are; energy, water quality, climate, and waste/contaminated land management. For each situation, there's an illustration of the twofold nature of technology's ecological implications (Yeoshua, 2005).

This manual also touches on the emergent association between innovation and environment quality, one which seems to strike both ways. It also highlights how technology is aiding to create policies that permit producers and customers to distinguish and internalize the ecological costs of technology, plus therefore spur innovation for cleaning up the environment.

As global economies continue facing massive challenges in the use of fossil energy, the lifeline of the industrial era, while preserving environmental quality. Even though U.S. power efficiency is way bigger than ever before, developments in the economy have assured growing energy usage. Whereas the average fuel capacity of new passenger vehicles has more than increased twofold since 1975, still environmental gains are gradually counterbalanced by the immense popularity of low-mileage lighter-duty trucks and sport utility vehicles (SUVs), increased miles traveled for a vehicle, and massive surge in vehicle ownership (Bharagava, 2020).

Even so, technology-propelled by economic, regulation, and environmental demands-has made possible striking reductions in vehicular discharges of volatile organic composites and carbon-monoxide per mile covered. Decreases by 70–80% since 1977 wouldn't have been likely without considerable innovations in, more notably, electronics.

The creation of sensors which can carefully calibrate energy usage to demand means both modern engines plus industrial motors may be run much more competently. Microcontrollers and virtual signal processors furthermore underpin a fresh generation of vehicle discharge sensors, now consuming up to 25% lesser power. Modern autos contain around 20 to 90 of these unique sensors for controlling their engines accurately (Bharagava, 2020).

1.1.1. Energy

Energy and Environment studies focus on the production, storage, and efficient use of energy and nature's resources and the evaluation of interactions between the ecosystem and energy technologies, for reasons of establishing clean and renewable power. Studies on the interactions between renewable energy innovations and the environment (nature's resources) is vital to comprehending and establishing sustainability.

This volume embodies the interactions between atmospheric and ecological conditions and the generation and use of clean, renewable energy. Improved solar, wind, and bio-energy systems, and the interrelations of these innovations with water resources, climate, and weather patterns are the key areas of focus. Two themes are of utmost importance when discussing energy and the environment, these are Sustainable Environmental Systems and Efficient and Clean Energy Systems. Research related to these topics explores current and new innovations to improve the efficiencies of different energy systems, while reducing the possible environmentally-disrupting side effects of these innovations.

1.1.2. Water Quality

Air quality and climatic changes while being the most pronounced are however not the only environmental issue that relates to energy usage and production. Manufacturing and vehicular emissions, especially of nitrogen oxides, may further be harmful to water quality. The nitrogen deposition functions as a fertilizer, promoting the emergence of algae living in lakes, rivers, or estuaries, forming eutrophic conditions which terminate submerged water vegetation. In some regions, like the Chesapeake Bay, there's risk of eutrophication that threatens commercial fishing and recreational endeavors (Johnson, 2006).

Even more concerning is the agricultural run-off of fertilizer, pesticides, and animal waste. Modern innovation and policy are presently starting to address run-off pollution, however, it's difficult to gauge, much less control, since it arises from widely distributed, "nonpoint" sources.

Within the last few years, nevertheless, the instruments of geographical information systems (GIS) that use remotely sensed information have provided new ways of identifying and observing these particular sources. Techniques combine land-usage data with hydrology, landscape, and soil data for making detailed, digital maps at highly fine scales and measuring the possibility for runoff (Johnson, 2006).

Remote sensing data on realistic farming activities, gathered by aircraft or satellites, may be merged with digital maps for providing more precise and timely assessments and evaluation of runoff. Though it might not be likely to trace every runoff to its primary source, it's progressively possible and affordable to trace a good part of it.

GIS technologies have also promoted accurate farm practices through real-time, electronic, and comprehensive information around plant health. Remote radars on harvesting equipment allow growers to differentiate among likes of crops for irrigation, and applying fertilizer or pesticides, therefore increasing crop harvests or reducing chemical use. Besides precision agriculture might have a positive future: information technology (IT) sales in the agricultural sector are presently comparable to the sales of common farm equipment (Cheremisinoff, 2002).

Remote sensing technology has furthermore started to enhance the proficiency of municipal water usage. Even in the U.S., water is calculated in a means of encouraging wasteful consumption. But the problem is found in many other nations as well, especially in developing countries, due to a lack of proper infrastructure for metering water use.

In some regions such as Buenos Aires, customers actually pay for their water depending on the scope of their homes or apartments, based on remote sensing data. Additionally, the city has lately updated its real-estate diagrams using remote sensor data. A number of hotels have been concealing as studio apartments, and being billed as so. Even though remote sensing hasn't replaced the necessity for metering, still fresh data has at least permitted the city to bill water more precisely.

Notwithstanding their potential, even the GIS and remote sensing technologies are "double-edged" in their ecological implications. These technologies raise certain privacy concerns, for example, that may cause polluters to hide or cloak their pollution activities, further obstructing contamination monitoring and cleanup (Cheremisinoff, 2002).

Various legal cases touching on constitutional protections against no warranty searches have equally been driven by the usage of aerial photography, particularly for assessing environmental compliance, and much recently polluters had tried to protect their actions from constant surveillance.

More recently, USA Midwest agricultural conglomerates have expressed alarm about the public accessibility of aerial imagery in case it's detailed sufficiently to disclose farming activities. These concerns may lead to restrictions on the usage of remote sensing for pollution assessments and regulatory enforcement.

1.1.3. Climate

Discussions around energy usage lead logically to the query of how it could be affecting the globe's climate. In America, the energy sector covers for roughly 85% of total greenhouse fumes releases, with energy-based carbon dioxide only being responsible for roughly 80%.

A good portion of U.S. greenhouse gas emissions come from using coal and petroleum for electricity production and transportation, correspondingly. But some newer innovations, such as energy cells and tiny, one-cycle gas turbines prompted by economic and ecological considerations, plus by innovation policy, provide considerable environmental benefits over traditional, massive, centralized power stations (FAO, 2000).

Localized generation by smaller facilities not only moderate transmission losses, but instead they also enhance air quality because they can easily be fueled through hydrogen and organic gas-which is cleaner than coal when measured on kilowatt/hour basis. In case fuel cells become broadly adopted in transportation, it means emissions may drop there too.

Adoption of these innovations may not be such a perfect solution, nevertheless, especially in power generation. Besides, some fuel cell technologies have been found to release carbon dioxide, a greenhouse gas. Additionally, small-scale plants which serve residential areas only or small businesses might be less capable of balancing the demand peaks than are bigger plants that serve both kinds of customers.

1.1.4. Waste and Contaminated Land Management

Some common waste and contaminated land management techniques in use today include: containment, extraction, pump-and-treatment, soil washing, air stripping, stabilization/solidification, precipitation, thermal desorption, biological remediation and vitrification. After a site has been assessed as polluted and requiring attention, additional research is necessary to examine the feasibility and possible success of proper remedial techniques. (Cheremisinoff, 2002)

When a site has been evaluated as contaminated and needing further attention, additional studies are necessary for investigating the feasibility and possible success of suitable remedial techniques. The forms of contaminants that present a substantial environmental hazard to either the environment or the suggested development should have been detected in the site investigation phase, but more comprehensive and extensive sampling is needed to address particular remedial needs of the contaminated land.

Based on the amount of contaminant source reduction needed to achieve the given goals, a variety of remedial options can further be regarded for their appropriateness and regulatory approval. This work normally involves using bench-scale models, and a review of environmental factors that may affect the process feasibility physically like odor release, space limitations and ground stability. Once a suitable remediation technique is arrived at between the different parties involved, detailed processes are established and the necessary funding, project management as well as sub-contract parties are detected and selected for reclaiming the contaminated land. (Ghosh, 2016)

1.1.5. The Information Revolution

The revolution in data technology promises significant economic changes, which are almost as impressive as those of industrial revolution itself. Virtual data storage, manipulating, and communication might not seem to contain environmental implications, however some examples recommend otherwise. Additionally, high-speed, top-bandwidth connection between homes and offices might allow for telecommuting; it may further degrade sprawl around metropolitan zones if personnel find it progressively practical to live much farther from their workplace.

If online shopping replaces tours to the mall, or occurs in addition to tours to the dentist or dry cleaners (trips which may have been merged with mall trips) will also determine the Internet's effect on auto travel. The packaging of e-commerce products for shipping might be more materials-plus energy-intensive compared to store-bought goods.

A number of controversial researches have even suggested increase in demand for electric power, propelled by new types of consumers like computer server warehouses, might have assisted in overloading electrical grids in northern California the previous summer. The overall effect of new data technologies focus on energy usage, land utilization, and travel has still to be fully studied.

From a different perspective, like an instrument for research or communication about environmental matters, the Internet seems to hold great promise. For studies, it provides digital bibliographic search engines, information archives and retrieval structures, swift exchange of research effects with distant colleagues, plus software for scientific structuring of complex environmental procedures. The internet has also incredibly grown the public's accessibility to and responsiveness of detailed environmental data.

1.1.6. Economic Incentives and Technology Innovation

Realizing the environmental potential of these and various other new innovations-that is, exploiting beneficial aspects of technology's double nature-relies in part on "having the prices right." Emerging technologies will be better employed to moderate environmental costs in case the costs are identified. For instance, if automobile costs reflect all the environmental rates of tailpipe emissions, it means auto makers shall have greater incentives to utilize new pollution control strategies in new vehicle models (OECD, 2011).

The "social costing" technique to environmental management has progressively come into its essence in the U. S. For example, tradable pollution permits-like for sulfur dioxide discharges from coal-based power plants-have formed financial motivations for electricity producers to adopt cleaner manufacturing processes.

The market-based approaches may be more affordable compared to traditional discharge restrictions or innovative measures, since companies that may reduce emissions most inexpensively reduce them by way more than they normally would-plus then sell their extra permits to companies that can't. At the same instance, the market-based techniques induce innovations through placing a price on discharges and reductions.

Additionally, the use of modern incentive-based techniques is growing not just locally in the U.S., but abroad as well. Global policy discussions concerning international climate change cover taxes on carbon discharges, and use of marketable licenses. Same approaches to getting the costs right in the management of water quality or waste, like in these examples, may possibly discourage environmentally risky uses of the resources apart from further encouraging the use of fresh technologies in handling them (OECD, 2011).

Information technologies (IT) in particular can help increase the scope and efficiency of incentive-based techniques, for roughly four reasons. For starters, improved remote sensor technologies are making incentive-oriented regulations, relying on emissions assessments, either to impose compliance or levy duties on pollution, providing increased practicality. Secondly, technological advances can help extend these methodologies even too much "smaller" polluters, probably including small-scale businesses and personal automobiles.

Thirdly, new information innovations are making it conceivable to fine-tune costs and regulatory systems-for instance, by permitting pricing to reflect the period of day, crowding, or atmospheric factors. Finally, for the case of global resource management, remote sensors from space satellites can make it simpler to assess environmental compliance across multiple countries (Skidmore, 2017).

Right from steel towns that dominated yesteryears to modern-day wired cities, the interaction of new technologies and environmental effects has rather been complicated. Technology shall continually be a double-side sword, however creative uses of new economic techniques to environmental management can help blunt its damaging edge and refine its capacity for the better.

1.2. WHAT IS ECO-FRIENDLY TECHNOLOGY?

Commonly referred to as clean technologies, green tech or environmental tech, eco-friendly technologies can help conserve the environment via energy efficiency, apart from reducing harmful wastes. Innovators in this field utilize the latest environmental sciences and green chemistry for reducing the harmful effects of human action on the globe (Figure 1.2).

Figure 1.2: Humans must preserve the earth for future generations.

Source: http://www.insectmist.com/services/eco-friendly-technologies/.

Ecotechnology is still at its earliest phases of development, however, many exciting technologies have certainly been made in aspects like renewable energies, water cleansing, and waste management, including in daily consumer products such as electronics or vehicles. It could be as tiny as a hand-based device or as extensive as the new technique of filtering greenhouse fumes out of the environment (Ghosh, 2016).

Eco-based technologies often include some of these attributes; recycled, recyclable, and bio-degradable content, renewable energy, plant-based materials, energy-efficiency, multi-functionality, low-impact manufacturing, minimization of pollution substances and reducing greenhouse gas discharges.

Lately, new advances in the area of green technologies have included incredibly efficient LED lights that may be used for making numerous other devices more eco-friendly, plus promising increased growth in less-known renewable products like algae oil. The company known as Agilyx is currently working on technology which may turn plastic rubbish into artificial crude oil, plus flexible thin-frame solar modules are a step closer to making the broader range of solar power applications possible (Ghosh, 2016).

In 2020, new measures for Energy Star models are already in the market and shaping the green revolution. Energy Star defines how efficient a model is in comparison to other models having the same size. More stars translate to greater efficiency in comparison to other similar-sized models. The U.S. Environmental Protection Agency (EPA) shall require TVs, cable cases and satellite cases to be 40% more efficient compared to conventional models for purposes of winning the label. Big techs like Google and Dell are taking prominent steps forward with environment-friendly technologies in projects varying from plant-based product packing to large wind farms.

Furthermore, Google invested roughly \$100 million for Shepherd's Flat wind program in Oregon, which serves an average number of 235,000 residences within the region as of 2020. It's the globe's largest wind farm supplying clean energy to the neighboring communities. The project helps to reduce 1.5 million metric tons of carbon discharges per year. Tech-based companies are primarily interested in the program since it shall be the first one to utilize direct-shaft turbines provided by GE (Maczulak, 2010).

However, wind energy isn't the only natural energy source that Google eyes; they also invested \$2 billion in a wind and solar energy project in 2019, making them the largest corporate investor in renewable electricity as of 2020, and also acquired a 49% stake in the photovoltaic (PV) ranch in Germany. These investments may help the technology giant power its unique energy-based operations more reliably. Google presently has around 1.6-megawatt solar connections of its own found at Mountain View, California.

Furthermore, Dell has brought some of its modern eco-friendly technologies much closer home-particularly the residences of consumers ordering the firm's computers. In 2020, they have a fresh sustainable packaging strategy which uses mushrooms for creating product cushioning for delivery. Naturally grown instead of manufactured, mushroom-based packaging is produced when farming waste products such as cotton hulls become pressed into patterns and then injected with mushroom-spawn (Hussain, 2020).

In around 5 to 10 days, the resultant packaging becomes ready for use. The mushroom-based packing is biodegradable, therefore making it a much greener option compared to commonly applied Styrofoam or polyethylene. Besides, eco-friendly technologies like these may continue having a big impact on tech industries, particularly when implemented by organizations with a big influence on customers (Maczulak, 2010).

1.3. EXAMPLES OF ECO-FRIENDLY PRODUCTS

Figure 1.3: Eco-friendly products are made from biodegradable or recyclable materials.

Source: https://medium.com/@ecoworkx/best-eco-friendly-products-to-use-in-2019-2d3d817cbdaf.

Many manufacturers are nowadays endeavoring to make their products as eco-friendly as possible. This is because of the fact a product may be regarded as eco-friendly, based on the measures which have been taken to produce the said item. You got the basic finished products like those which are regarded as bio-degradable. These may be items produced from paper that can easily disintegrate.

Likewise, there are items made from recycled products. They may be regarded as "up cycled" items. These may be regarded as eco-friendly since the products being re-used would otherwise finish up in the landfill. In case you need examples, then Google up cycled programs and the results shall blow up your mind (Bharagava, 2020) (Figure 1.3).

Furthermore, products that are normally labeled cruelty-free may fit easily into the eco-based definitions. They are products which don't perform animal testing. Some examples of these kinds of items from the information available, shall be Wet and Wild and Perfectly Posh beauty items. You may check for what's known as "leaping bunny" emblem on products to determine for sure whether the said products are cruelty-free or otherwise.

Furthermore, you also have items which are regarded as eco-friendly just by their supply sequence process. The items are normally branded "sustainably sourced" or "organic," taking into account the environment during the entire production process. Items made from hemp may very well rank within this category (Bharagava, 2020).

Similarly, there are third-party certifications which can show you if an establishment is actually attempting to apply eco-friendly activities into their manufacturing process. Apart from a leaping bunny certificate, another possible consideration may be to see whether they fit the established guidelines for certification as a Benefits Corporation. It means that they must complete the procedure to take appropriate steps to not merely consider profits, but also humans and the planet also throughout their corporate model.

Nowadays, there are eco-friendly straws which are reusable, biodegradable, and/or compostable, and have stylish trend, just like the interiors of restaurants, bars, and households have glass straws that add a touch of style to dining rooms. The wide variety of green options accessible when products are considered, has been spurred by increasing demand for sustainable straws market.

The eco-friendly straws are either made up of bio-based materials such as paper, or reusable materials which includes food-grade silicone, metals or glass. The environment-friendly straws are produced in various designs such as strip, star-shaped, and polka dot among other printed papers. These eco-friendly straws provide manufacturers with affordable solutions. Producers are also offering varying diameter choices in recyclable straws according to brand owner standards, and vendors demand (Bharagava, 2020).

In brief, there are various factors that may make a product to be eco-friendly. These include:

- Bio-degradable or recycled packing material;
- Minimal packaging material;
- Creating a product without using toxic compounds (consider pesticides used on apples or cotton flowing off into waterways, poisoning individuals, animals, and crops);
- Highly sustainable raw products (consider natural rubber vs. synthetic options, artificial cotton vs. natural cotton, and virgin polyester in comparison to recycled polyester); and
- Products made in sustainable factories (this may be a factory which uses minimal water, renewable power, filters, and cleanses its wastewater, and so on.).

Generally, there are various manufacturing procedures that affect the health of individuals, animals, and the environment at large. Consider countries which are prolific manufacturers for example. Environmental-friendly products diminish that damage, eventually striving for the perfect goal which is having a neutral impact on the earth, or more remarkably, a positive effect.

1.3.1. Eco-Friendly Straws

Some common examples include bamboo straws, recyclable straws, metallic and glass straws which are gradually replacing plastic straws, since the latter creates a large capacity of plastic waste (Figure 1.4).

Figure 1.4: Eco-friendly straws industry is expected to experience significant growth based on the amount of strict regulations imposed on plastic usage.

Source: https://www.shutterstock.com/image-photo/ecological-bamboo-straw-tube-drinking-water-1109341754.

Roughly 500 million artificial straws are thrown away daily in the USA, however starting from 2015 up to now, more cafes and hotel chains are choosing eco-friendly straws. Global use of eco-friendly items is ultimately affecting trends and designs in the food/beverages industry. As from plastics ban, promoting eco-tourism and using biodegradable products, including various initiatives that have been happening in different states to make them eco-friendly.

Overall, while plastic is stronger, paper is more flexible. It means that if the paper straw made its way into oceans, it's much less probable to be damaging to marine ecosystem. It's also likely to biodegrade, or rather disintegrate, over time—while plastic straws will endure forever. For the shortcomings, natural paper straws come with more greenhouse gas emissions during production compared to their plastic counterparts.

1.3.1.1. Bamboo-Based and Paper Straws

Single-usage plastic straws are gradually going out of fashion as more individuals begin recognizing their destructive, wasteful, and eco-destructive properties. Many leading food joints are now implementing compostable straws, plus 2019 onwards shall scc a considcrable jump in the usage of these straws.

Some experts have also proposed total haulage, distribution, and the transportation of perishable products to be electrical, so as to avoid the use of plastic sealants and packages that are bad for the environment. Basically, food deliveries may be in hermetical sealants and where necessary cooled containers, whereby delivery personnel may allow for the products to be moved from those enclosures onto refrigerated utilities and cupboards.

Furthermore, transport containers may be returned for cleaning and re-use, therefore saving on unnecessary wrapping and providing greater conservation. Eventually, the result is a quiet, hygienic, and efficient way of servicing multiple recipients and providing a touch of personal service to transportation.

In essence, eco-friendly products can be defined as just about any product that's made from stainless steel, recycled plastics, glass, and organic materials such as bamboo and organic cotton, hemp-based products, or other fabric oriented products apart from cheap plastics.

1.3.2. Bamboo Toothbrushes

Several millions of synthetic toothbrushes add up to the already overloaded waste disposal systems each month. Many companies have begun working towards creating bamboo toothbrushes plus have conducted many campaigns towards spreading awareness concerning the presence of this green alternative (Figure 1.5).

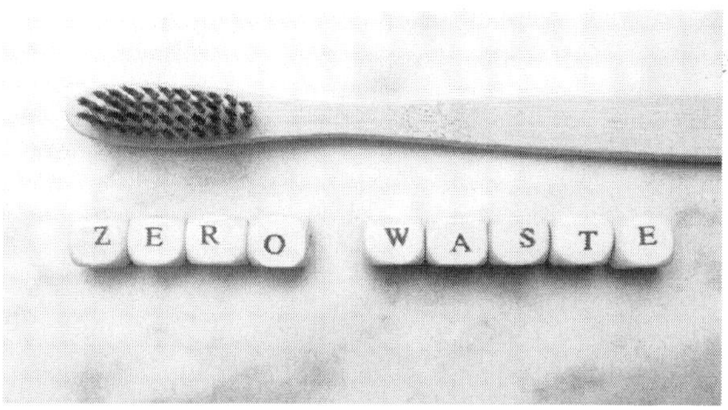

Figure 1.5: Bamboo toothbrushes.

Source: https://www.istockphoto.com/photo/eco-friendly-bamboo-toothbrush-gm1139378374-304519456/.

1.3.3. Bio Bags

The mission to plug the explosion of plastic bags around the world has motivated businesspersons to develop greener options. Currently, there are multi-purpose waste bags prepared from compostable materials, such as organic starch or potato, making them very biodegradable (Figure 1.6).

Figure 1.6: Biodegradable bag.

Source: https://biobagworld.com.au/product/bio-bag-carry-bags-small/.

1.4. THE ADVANTAGES OF ECO-FRIENDLY PRODUCTS

A rapid increase in the global population, coupled with an increase in industrialization and the subsequent increase in pollution, disasters, and diseases has led to an increasing embrace of the concept of sustainable development. Communities have become aware of how "business-as-usual" has impacted human and ecosystem health. An increased awareness of the circularity of the impacts that humans inflict on the natural environment and the resultant negative impacts on our health and well-being has led to recognition of the importance of embracing an eco-friendly lifestyle.

It's like everywhere we go, there's someone who has gone green, and individuals talk about the way it influences our planet in a massive way. Going ecofriendly means living a lifestyle which isn't just more valuable for persons who practice it, but also for their environs.

Eco-friendly items are so-called since they're less harmful for the environment. For instance, going green means equally minimizing your carbon footprint. (Maczulak, 2010).

Sustainable products provide benefits for people not only economically, but socially and environmentally as well, meaning they preserve public health generally. The more individuals resolve to go green, generally the better it gets for everybody around them. When the individuals become happy, the entire society also functions optimally, and vice versa.

When using products which are all-natural, it's way better for people's wellbeing. It should also be noted that they last much longer (organic materials are more robust and have better quality). It's related to everything including food, clothing, and cosmetics among others. For instance, with regards to clothes only, nowadays there are various amazing brands which make their products from materials with the best quality while striving to be environmental-friendly.

Nowadays, it's possible to find good quality undergarments, jeans, even various baby clothes. Besides, by teaching youngsters to consider organic options when they're still young, equally we teach them to become more concerned and mindful of the environment surrounding them. We likewise want our children to always select the healthiest preference (Maczulak, 2010).

Being eco-friendly is rather rewarding, since we get that great feeling that we're doing something meaningful in life. Pollution and contamination is also greatly reduced. In case this were the only reason why being eco-friendly is good, then it's still a pretty impressive reason.

It's correct to say because of some products that we use, reduction of pollution and contamination is possible. Pollution is a major problem which we deal with today plus it threatens to endanger our future. However, if more individuals go green, then the less we shall have to stay in fear of large potential disasters.

Besides, our kids can have a brighter future. There's a saying which mentions like this: we don't inherit the world from our ancestors, but rather we borrow it safely from our kids. It's why we should do everything in our control to treat the environment right, such that we may save it for the children who come after us (Maczulak, 2010).

Not only saving it, but making it appear as decent as possible. Green items are certainly one of the various ways to allow our children enjoy a brighter future. They shall appreciate it considerably, plus we'll feel good of ourselves since we left behind something wonderful for the succeeding generations (Bharagava, 2020).

Furthermore, through using organic and natural green products, it's possible to save lots of cash in the long haul. This may be seen when talking about stuffs or objects we need at home, like lightning and some devices which save energy. The U.S Energy Department says that when properties are all green-improved, then America would use roughly $20 billion less power each single year. Not to mention there are lots of discounts when we make our houses a greener one.

Everyone has a responsibility of doing things that are best for our general well-being, but also taking care of everybody and everything surrounding us. Among the greatest attributes about life and existing as a human being is that we're in the world together. People rely on the society, while in turn the society relies on its people. Going green simply means doing more for the society, ecosystem, and planet with regards to conservation.

When going to the store for purposes of buying a new shampoo bottle, or some other cleaning materials for the house, it means the environment might not actually be the first aspect on our minds since we contemplate which products to buy. Nevertheless, with Earth Day just about the corner, it's important to contemplate on the benefits of choosing eco-friendly items as opposed to other products (Yeoshua, 2005).

Prior to considering these factors, it's important to define exactly what it means when we say "eco-friendly." To be eco-friendly simply means that the product doesn't damage the ecosystem, both during production of the object and using it. These items, according to (EPA), "offer benefits for people economically, environmentally, and socially, while conserving the public health." Individuals who use these products are assisting to lessen pollution and impurity of the water, soil, air, and so on. Here are some reasons why moving to eco-friendly items may be helpful to you:

They Offer a Healthier Lifestyle: Not just are these green items healthier for the ecosystem, they are also much safer for everyone. There are many products which contain toxic chemicals that put people at risk of contracting illnesses and ailments. Environmental-friendly products remove all of these hazards by using natural items and avoiding unnecessary additives we know little about.

Environmental-Friendly Products Utilize Biodegradable Packaging: Plastics are usually the go-to product used for packaging, but are very destructive to the environment. Environmental friendly manufacturers make items using packaging that's made from recycled materials which can easily decompose… plus a rather fun fact, many manufacturers implant seeds in wrappers so they can end up nurturing a plant if they're transferred to the landfill.

Using These Items won't Conflict with Personal Morals: Not just are these items healthy for the ecosystem with regards to being biodegradable, instead they also don't test their produces on animals or utilize any animal by-products in manufacturing. All of these negative aspects which often come with substandard manufacturing of bigger companies is totally eliminated.

Keeping the Environment Happy and Healthy: The air we inhale and land we stay on is often harmed by pollutants and toxins; the corporations that manufacture eco-friendly items steer free of these factors by using items like solar energy to operate their plants. Beware that even though a company might advertise their items as being "eco-friendly," still they may not actually be eco-friendly during their manufacturing process; thus, do your own research on the firm you're buying from.

Prevents Environmental Degradation: It's among the main reasons our world is becoming degraded nowadays. This is possibly related to falling down of trees, regulating supplies, exploiting natural resources and so on. Even water consumed today in many cities around the world isn't fit for consumption, since most of the industries overlook toxic products and waste items discarded into pure or freshwaters. Therefore, they affect water purity and further kill marine life. Going green is the best way to conserve nature and leave a legacy for future generations.

1.5. HOW CAN TECHNOLOGY SAVE THE ENVIRONMENT?

Modern technology is changing much faster than people can keep up, these new smart technologies are changing how we stay and work. A limitless stream of useful new apps adding to our smartphones, including the emergence of artificial intelligence (AI) generally improve how we interact with information. Apart from making our lives much easier, this also helps us to live a greener, better lives (Figure 1.7).

Figure 1.7: Technology shows great promise in promoting conservation efforts.

Source: https://www.fibre2fashion.com/industry-article/7224/environmental-strategies-adopted-by-companies-to-achieve-sustainability.

Innovation and the environment might actually be more deeply linked than people may think. Our regular technological systems have the capacity to damage the environment, however, if we utilize them wisely and create sustainable concepts, they may also help to solve the environmental problems that we've created. Technology might not be the only answer to climatic change, however, it could be an essential aspect of the solution. Below are some common areas where technology can positively impact the environment:

Renewable Energy: Among the most renowned and significant eco-friendly tech developments in recent periods have been in clean-energy sector. Some renewable energy sources like sunlight, wind, and hydroelectric energy have become more commonplace and affordable. The technology is progressing also. Technological advances have made renewable energies become more resourceful. Products such as Tesla's solar panels, resembling regular roofing tiles have led them to become easily available and alluring to consumers.

Going Digital: Companies and individuals are increasingly using lesser paper than in the past years, all thanks to workstations, smartphones, and cloud-based storage. Several companies nowadays send bills, newsletters plus other communications through email instead of regular mails. Folders are kept in the cloud system, which is readily accessible from anywhere, instead of physical filing cabinets. Individuals are more regularly taking notes and handling their schedules through smartphones instead of physical notepads or calendars. Altogether this contributes to a "new standard" that has lowered the overdependence on the globe's tree population, therefore helping to diminish deforestation.

Environmental Monitoring: Technology is widely being applied to ensure environment-based laws and procedures are being adhered to. In regions where poachers regularly hunt endangered animals, environmentalists have begun applying drones to have a closer eye on the region. The drones serve as security cameras which can cover immense distances, therefore making it harder for those doing illegal hunting to get away after breaking the law.

Governments also apply technology to monitor if companies are following conservation guidelines. Geo-spatial air pollution measuring structures draw in air from around gas/oil facilities for test the average ratio of pollutants present. Additionally, Environmental Simulation Tests ensure that new products and innovations are up to high quality standards before hitting the market, and beginning to affect the planet.

The Sharing Economy: The so-called sharing economy has arisen as a new development that moderates our environmental impact by allowing people to buy less and use lesser materials. Businesses like Uber have made it easier for us to move around without even owning a vehicle. Airbnb allows homeowners and travelers to utilize spaces which aren't being used, thereby consolidating the region and resources required to travel. Besides, services like Netflix have lessened the amount of DVD rent outs in favor of digital streaming, therefore reducing the requirement for postage, moving items and developing physical models of movies.

Electric Cars: Even though ridesharing businesses allow people to use lesser vehicles, those cars are also getting more sustainable. The amount of electric cars used today has risen progressively over the last few years. These hybrid cars have furthermore become more commonplace. The general price of electric vehicles is dropping also, because of technical advancements in the rates of the batteries required to operate them. As a matter of fact, experts envisage that they'll become more affordable to own compared to traditional cars by 2022.

Smarter Homes: The household is among the largest energy users; however new technologies are constantly helping to alter that. Instruments like smart thermostats or motion-activated lighting are making it simpler to only use energy when you actually need it, therefore saving both cash and energy. These devices may be utilized in both private residences and in companies, plus they have the capacity to minimize energy usage considerably. Whereas technology and the ecosystem may look like two very different aspects, nature, and some of our most advanced devices are greatly interconnected than people may think.

Nature has been inspiring technology as well. It provides us a reference point from which it's possible to innovate, and teach people what can and won't work. Furthermore, technology has the ability to transform nature- either for good or bad. One lesson that we're only learning now from nature is we must use innovation in a sustainable manner. Technology can help people to use much less resources, power our lives in a different way and be highly aware of our impact on the environment. Technology may form part of the answer or an aspect of the problem. People are finding more ingenious ways of ensuring technology helps us instead of harming, a trend that we must ensure sticks around.

Saving Energy: It might be surprising to a few people that smart products are helping people to not just simplify their lives, but also saving the environment. The smart home products conserve energy through using AI to shut off power to implements, not in use. Additionally, smart products like refrigerators may instantly send you an alert email in case the door is partly open, or if items stored in the fridge are about to expire.

Smart home implements may also optimize the thermostat or power usage, so that you don't consume unnecessary power when you're not present at home. Besides, thermostats, and digitized bulbs may be programmed for turning on or off at given periods, and some may automatically adjust through detecting activities happening in the room. This is a great means to know if somebody in the house is staying up late, or the light bulb is being used irresponsibly.

Conserving Resources: Just like your appliances may allow you to know of any energy-wastage, they may also email you in case they have any part which is about to be worn out. It doesn't just save you cash on replacement of expensive appliances, but further permits you to fix issues before your primary instrument ends up in some landfill. Nowadays, there are smart sprinkler units that help to regulate the amount of water you're using, by regulating based on weather, moderating not just your water bills but also impacts on the communal water supply system.

Green-buildings which are built to ideally harness and apply natural lighting can drastically diminish lighting costs plus reduce the usage of coal required to power the lighting. Because electric power accounts for nearly 93% of all coal usage in the U.S, it provides immense possibility for reducing carbon footprint.

Eliminating Waste: Even though we haven't yet become the paperless system of early predictions, still screen technology provides a great alternative to traditional pen and paper for interactions and projects, plus an alternative to printable materials through a growing media system that comprises of eBooks and audiobooks. The streaming of media services is also reducing the requirement for producing DVDs, including all the wasteful packages that go along with it.

Innovation is also being used to help avoid food waste, therefore working to ensure food stays out of landfills through keeping improved track of the garbage at all levels, ranging from the garden to the table. Having up to 1/3 of the globe's food resources discarded, data collection innovation has the ability to greatly diminish that figure.

Information collection and analysis may further help charities to connect with partners such as grocery stores or restaurants to steer surplus products to individuals who need them. Technologies in bio-fuel digesters are turning food waste to usage in developing new energy. At the home also, smart refrigerators are assisting families to keep greater track of their varied food resources, such as expiration dates, with some even providing recipes making use of food that's about to go stale.

Makes Users Smarter: Our children are nowadays learning in smarter classes that are developed around creating better interactions and active involvement, often integrating smart projectors, digital whiteboards, and custom digital implements. Schools are developing or upgrading their systems to classrooms which are better at preserving energy and developing a more comfortable education space, including smart climate management and lighting that resembles natural daylight, enhancing alertness, and offering the opportunity for teaching students more about preservation and environmental resources.

Besides, technology is further helping people to work smarter, thus providing better methods for analyzing data, improving workflow, streamlining supply chains, detecting problems faster, and improving production processes. Lesser garbage in the system ultimately means more sustainable activities broadly.

Promotes Improved Health: The ways that technology is assisting people to get healthier are becoming innumerable, beginning with basic phone apps which can help individuals meditate more, motivate them to take more water or eat better, and promote exercise. For homes, digital bulbs may change or lessen blue light emission during certain periods, which may help provide better sleep patterns and safeguard your vision as well.

Continuous improvements in medical technology, plus the introduction of AI into modern medicine, is allowing for improved diagnoses and treatment opportunities, accuracy of treatments like radiation therapy, and recognition of early degeneration patterns. These emergent technologies are providing physicians and caregivers more targeted data and allowing them to deliver better results in treatment of patients. Healthy people have greater capacity to work in a smart mode, shop, play, and think smarter with regards to the environment-something that can only cause better ways of conserving the planet.

The earth as we know it today is in danger. People are either causing or contending with ocean acidification, deforestation, and rising temperatures, just to mention but a few of the negative legacies humans are leaving behind. While the facts are complicated the trends aren't readily disputed. But yet, the subject of global environmental risk caused by pollution is being tackled these days through products like green technology.

1.6. TECHNOLOGIES CONTRIBUTING TO THE GREEN REVOLUTION

Figure 1.8: Preserving the environment allows for more biodiversity.

Source: https://www.conserve-energy-future.com/methods-and-importance-of-environmental-conservation.php.

Solar Glass: Imagine if every window found in a skyscraper building could produce energy. This is the potential that solar glass has, an emergent technology that's receiving lots of buzz with regards to design and environment preservation circles. Same as it sounds, the solar glass is a fittingly transparent window material however also traps the sun's power and transforms it into electricity (Figure 1.8).

However, the largest hurdle has been proficiency. High-delivery solar cells can attain 25% efficiency or more, but retaining transparency means forgoing the efficiency whereby light is changed into electricity. Even so, scientists from University of Michigan have embarked on creating a solar glass item offering 15% efficiency, and increasing while allowing a complete 50% of light to pass through. Based on solid projections, in a state like Michigan alone there are 5-to-7 billion sq. meters of functional window space existing today, just enough for powering a full roughly 40% of the U. S. energy needs, all this with only a solar-glass product.

Graphene: This material is stronger than steel, but thinner than paper, and even more conductive compared to copper, indeed graphene is a miraculous material. It's an ultra-thin film of graphite which was originally discovered in 2004 by the University of Manchester scholars. Graphene is now the topic of concerted research and speculation efforts, with many people predicting it would be the next big thing after iron, steel, bronze, and silicon in promoting the cultural and technology evolutions of our human species.

Measuring only one-atom dense, graphene is a flexible, clear, and impressively conductive product, therefore making it appropriate to a massive range of planet-conservation applications. They include water filtering, superconductors capable of transmitting energy across large distances with reduced loss, and PV usage, just to mention a few. Through vastly improving efficiencies over existing materials, graphene might prove to be the keystone in eco-conservation rebirth.

Plant-based plastic*:* Plant-derived plastics which are biodegraded provide one palatable solution, since they could, theoretically, replace most of the plastic items already in circulation. One company known as Avani Eco has positively been manufacturing bio-plastic made from cassava as of 2014. This is not only a sustainable practice, but is also expected to become a business opportunity over the next coming years.

CHAPTER 2

Renewable Energy Technologies and the Environment

CONTENTS

2.1. INTRODUCTION

The use of renewable energy is booming, mainly because of the high level of innovation that has helped in bringing down costs in addition to promoting a future of clean energy. In the United States, the generation of wind and solar energy has undergone integration into the electricity grid that is being used nationally, with no compromise to its reliability.

This means that they have been able to start the process of displacing "dirty" fossil fuels that are currently being used in the power sector, therefore having the ability to offer the advantages of lower carbon emissions as well as other forms of pollution. It is important to note that not all energy sources that are said to be renewable are good for the environment. There are some that create difficult tradeoffs, such as large hydroelectric dams and biomass, especially when you consider the impact it has on climate change, wildlife, and other issues. There are some aspects that need to be understood when it comes to the types of renewable energy sources, as well as how these technologies can be used to replace dirty fossil fuels. This will be explained in this write up (Figure 2.1).

Figure 2.1: The main renewable energy technologies are bioenergy (biomass and biofuels), wind energy, solar energy, hydropower, geothermal, and ocean.

Source: https://www.eurekalert.org/pub_releases/2019-02/nu-ntf013119. PHP.

Renewable energy, also called clean energy, is energy that is derived from natural processes or sources that can be replenished at a constant rate. For example, the wind keeps blowing and the sun keeps shining, despite the fact that their availability is all dependent on weather and time (Sayigh, 2012).

Most of the time, renewable energy is considered as new technology, however, the ability to harness power from nature has been available for a long time, where it has been used for lighting, transportation, heating, and more. For the longest time, wind has been used to power windmills for grinding maize and even boats for sailing the seas. People have also used the sun many times to kindle fires that last long through the evening, and even just for the provision of warmth. However, over the past 400 years or more, we started to turn to energy sources that were much cheaper and dirtier, such as fracked gas and coal.

Due to the fact that we have been able to create methods of capturing and retaining solar and wind energy, that are less expensive and much more innovative, renewable energy is becoming very important and necessary. In the United States, it accounts for an eighth of the total power generated. There are even some communities in the rural areas that entirely depend on renewable energy for lighting and heating. With the continuous growth of renewable energy, there is a need to modernize electricity grids to make them better integrated across all regions, more secure, and much smarter (Sayigh, 2012).

Some of the nonrenewable energy sources include fossil fuels like coal, gas, and oil. These sources can only be found in amounts that are limited, and when they are used, they take longer time periods to be replenished. For example, gas that is pumped into a car is from a resource that is finite, and this resource has undergone refining from crude oil, which has been in the ground since prehistoric times.

In addition, nonrenewable sources are only found in certain countries in the world, meaning that some nations have more than others. This is in comparison to wind and sunshine which can be accessed by every country. Prioritization of nonrenewable energy can help in the improvement of a country's national security through the reduction of the country's reliance on these energy sources from nations that have them in plenty (Pimentel, 2008).

There are many nonrenewable sources of energy that can harm human health or even endanger the environment. For example, coal power plants foul the air, the fracking technology can cause water pollution and earthquakes, etc.

There are different forms of renewable energy sources. These include:

- **Solar Energy:** solar energy has been harnessed by humans for a very long time for drying foods, staying warm and growing crops. In fact, research has shown that the amount of energy that falls from the sun in one hour is much more than that which is used by us on earth in one year. There are many different ways in which we use the sun as of today: for powering devices, warming water, or heating businesses and homes (Figure 2.2).

Figure 2.2: Solar energy is mainly derived from the sun, and this energy can be captured through the use of various technologies; however, the main one is solar panels.

Source: https://news.energysage.com/what-is-solar-energy/.

- **Wind Energy:** windmills were used a lot back in the day, and we've come a long way since then. These days, engineers have been able to construct turbines that can go as high as a skyscraper, and some almost as wide. The blades in the turbine are turned by wind energy which in turn ensures that an electric generator is fed, thereafter producing electricity. Wind energy has actually become one of the cheapest sources of energy all over the world. It however is affected by weather.
- **Hydroelectric Power:** n the United States, this has been the largest renewable source of energy for electricity, however, wind energy is pushing very close to taking the lead. This type of energy is reliant on water—generally, water that is moving at a high speed in a large river, or water that is descending from a high point at a high speed—working to convert the water's force into electricity through spinning the turbine blades in the generator. Internationally and nationally, mega-dams or large hydroelectric plants are more than often considered to be energy that is nonrenewable. These plants normally cause the diversion or reduction of natural flows, thus leading to restricted access for human and animal populations that normally need these rivers to survive. However, plants that are smaller in size and also carefully managed do not cause much damage to the environment because they only divert a small amount of water flow.

- **Biomass Energy:** this is organic material that is created from animals and plants. It is inclusive of trees, waste wood, and crops. When we burn biomass, the chemical energy is then released as heat which can be used for the generation of electricity with the use of a steam turbine. Most of the time, we mistakenly describe biomass as a greener alternative to fossil fuels such as coal, or a clean renewable fuel for the production of electricity. However, this has been debunked by recent science which has been able to prove that many types of biomass, mostly that which we acquire from forests, tend to lead to the production of higher levels of carbon emissions in comparison to fossil fuels. These also cause negative effects on biodiversity. However, under the right circumstances, there are some forms of biomass energy that can provide an option that has lower carbon emissions. For example, chips, and sawdust from sawmills can be used for the production of low-carbon energy instead of leaving them to decompose and release carbon.
- **Geothermal Energy:** the core of the earth is approximately as hot as the surface of the sun. This is mainly due to the fact that the radioactive particles that are found in the rocks at the earth core take a long time to decompose. When we drill deep wells, we are able to access water from underground that is very hot as a hydrothermal resource, which can then be pumped into a turbine for the creation of electricity. Generally, geothermal plants produce low levels of emissions if the water and steam that they use are pumped back into the reservoir. It is possible to build plants that do not require underground reservoirs, however, these have brought about concerns when it comes to increasing the risk of an earthquake in areas that are already geothermal hot spots.
- **Marine Energy:** The wave and tidal energy are still being developed, however, the moon's gravity will always rule the ocean, making it much easier to harness the power of the ocean. There are some approaches of tidal energy that may harm wildlife, such as tidal barrages which are found in lagoons or ocean bays, and they work much like dams. Similar to tidal power, power generated from waves is also highly reliant on devices that have been anchored on or just below the surface of the water, or dam-like structures (Figure 2.3).

Figure 2.3: Marine power, also referred to as marine energy, is a term used to refer to the energy that ocean temperatures differences, salinity, tides, and ocean waves carry. The water movement in the oceans all over the world create a large store of energy in motion.

Source: https://www.irena.org/ocean.

These are some of the forms of renewable energy that we will continue to discuss in this topic.

2.2. BIOMASS ENERGY

As explained above, this is a simple term used to refer to all organic materials that are acquired from plants, crops, and trees. It is, therefore, true to say that there is a level of diversity when it comes to biomass sources, including crops that are grown specifically for the production of electricity, fuels, and heat, forestry, and agricultural residues, as well as organic waste streams (Pimentel, 2008).

Biomass provides a high contribution to the energy supply globally. It mostly contributes to energy consumption in developing countries, contributing approximately a third to a fifth of the total energy consumption. This is in comparison to 3% in countries that are industrialized. Traditionally, the type of biomass that is mainly used in developing countries is firewood for heating and cooking. It is important to note that some of the traditional uses of firewood are not sustainable due to the fact that it may lead to certain issues, such as causing outdoor and indoor pollution, deprive the soil of nutrients that it may require, and even cause poor health.

Biomass may also lead to increased greenhouse gas emissions which causes negative effects on the ecosystem. There is an estimated 7 exajoules use of biomass use every year for the production of biofuels, steam, and electricity in the modern setting. This is all commercial use, where it is used for commercial purposes. This means that the traditional use of biomass each year remains high at an approximate 38 exajoules (Pimentel, 2008).

Ever since early in the 90s, biomass has increasingly gained an interest globally. When it is produced in a sustainable manner, it is carbon neutral. It has a relatively even geographic distribution. When used in a sustainable manner, it is possible to clean and convenient to use modern energy carriers. This means that it is possible to use it to make a large contribution to the development of rural areas (Figure 2.4).

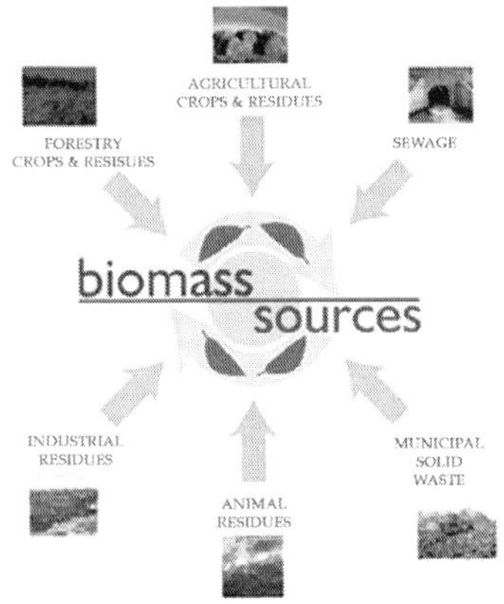

Figure 2.4: Commonly available sources of biomass for the production of energy. These sources provide energy that can either be burned or undergo conversion into electricity.

Source: https://www.pinterest.com/pin/464574517782331875/.

Biomass has costs that are attractive, making it a source of energy that is promising in many areas. The application of modern and commercial biomass energy is growing all over the world due to the development of different technologies that are able to ensure the conversion of biomass into modern energy carriers (Nelson, 2013).

Biomass energy has a resource potential that is much larger than the current energy consumption of the world. However, due to the fact that the conversion efficiency of solar to biomass energy is low, there is a need for large areas for the production of sustainable amounts of modern energy carriers. Due to the modernization of agriculture to standards that are reasonable, and also the need for the preservation and improvement of global natural areas, there maybe 800 to 1,500 million hectares of land available for the production of biomass energy. This is all inclusive of lands that are unproductive and degraded, as well as excess agricultural lands. How available land is for energy plantation is mostly dependent on the required food supply and also on the possibility for the intensification of sustainable agricultural production.

There are many different conversion routes for the production of fuels, electricity, and/or heat. These are explained below:

Heat Production*:* In countries that are developing, the introduction and development of better stoves for heating and cooking can provide huge impacts on the use of biomass. There is a widespread use of domestic biomass-fired systems that are used for heating in countries that have colder climates such as Germany, Austria, and Scandinavia. Heating systems that have been improved are now automated, using standard fuel (such as pellets) and having catalytic gas cleaning. There is a considerable advantage of this over open fireplaces, with the improved domestic heaters providing an efficiency of greater than 70%, as well as producing less emissions to the atmosphere. Presently, the estimated heat-generating capacity of thermal energy is approximately 200 gigawatts.

Production of Electricity: There are many commercial applications of the combustion of biomass for the production of electricity in many different areas, with an estimated total installation capacity of approximately 40 gigawatts of electricity. Applying advanced gas cleaning and fluid bed combustion helps in efficiently producing heat and electricity from biomass. Most of the time, electricity is produced together with steam or heat in Sweden and Denmark. Sawmill factories in Southeast Asia, mainly Thailand, Malaysia, and Indonesia, use congregation systems that utilize wood waste that is retrieved from factories. Co-combustion systems, such as those that combine coal and natural gas with biomass, are created in such countries, such as Denmark, that have the advantages of economies of scale that are greater, and lower risks when it comes to fuel supply. In coal-fired power plants, co-combustion of biomass is a popular method of increasing the power generation capacity based on biomass with less investment. In comparison to coal-based power production, another advantage of using biomass is that it helps in producing higher efficiencies as well as less nitrogen oxide and sulfur dioxide emissions.

Large Gasification: Gasification technology works by taking biomass and converting it into fuel gas, which is thereafter cleaned before it is combusted in a gas turbine. Biomass integrated gasification/combined-cycle systems are able to ensure the combination of high electrical efficiency and flexible fuel characteristics. In certain countries, there are certain demonstration projects underway for certain concepts of gasification (Figure 2.5).

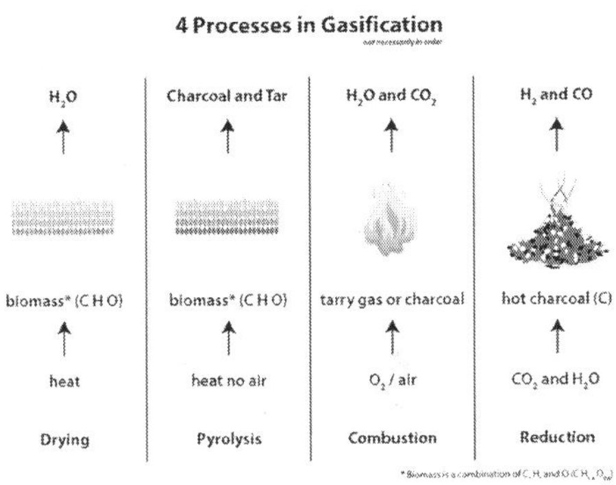

Figure 2.5: Above are the four biomass gasification processes. They include: Drying, pyrolysis, combustion, and reduction. All these processes lead to the production of carbon emissions.

Source: http://www.infiniteenergyindia.com/biomass-gasifiers.html.

There was a project that was initiated in Brazil between 2013 and 2017 which was able to demonstrate a 30 MW electric Biomass integrated gasification/combined cycle system unit fired with cultivated eucalyptus. Sweden was able to create a Biomass integrated gasification/combined cycle system unit that was based on pressurized gasification, and as of today, this has had thousands of hours of operational experience. In Yorkshire, United Kingdom, they were able to commission an atmospheric Biomass integrated gasification/combined cycle system. Another option is the co-gasification of biomass and this is being applied in Europe and the United States. For fuel gas produced using biomass gasification, there has been development of an interesting alternative, which involves using it in either new or existing natural gas-fired combined cycles. This alternative makes it possible for economies of scale to come with a fuel supply that is safe. This alternative has however not yet been demonstrated, but there is research being done on getting it there (Nelson, 2013).

Small Gasification: On the market, it is now possible to find small gasifiers that have been coupled to gasoline or diesel engines commercially. These have an approximate electrical efficiency of 15 to 25%, and are mainly created for systems that have 100 to 200 kilowatts of electricity. Their application in high numbers is however blocked due to their need for careful operation and gas cleaning, as well as their high costs. However, in China, Indonesia, and rural India, there are some systems that are successfully being used.

Biogas Production: There has been commercial application and demonstration of anaerobic digestion of biomass with success in many different situations and for many different feedstock's, such as sludge's, manure, organic industrial waste, and organic domestic waste. The development of large advanced systems in mainly for industrial waste streams that are wet, and these are being used in many countries. In India, animal, and other waste are mainly being used to produce biogas which is then used for generating power, lighting, and cooking. Anaerobic digestion tends to provide a low overall efficiency when it comes to electric conversion, and it mainly works better specifically for wet biomass materials. There is a high risk of increased atmospheric methane emissions mainly contributed to by landfills. Many times, collecting landfill gas and converting it to electricity with the use of gas engines has proven to be profitable, and there is now a much more widespread use of these systems.

Producing Gaseous and Liquid Fuels from Biomass (bio-crude and bio-oil): In the absence of Oxygen, and at temperatures of approximately 500 degrees Celsius, pyrolysis is able to create solid (charcoal), gaseous, and liquid (bio-oil) fractions from biomass. When using fast pyrolysis techniques, which is also referred to as flash pyrolysis, it is possible to obtain a liquid fraction of up to 70% of the thermal biomass input. Bio-oil is acidic and corrosive, and it contains approximately 40 weight-percent. It is possible to upgrade this oil to lower its oxygen content, however, this has energy and economic penalties. This technology is still being developed. Shell was able to develop Hydrothermal upgrading which works by converting biomass in water at moderate temperatures and at high pressure into bio crude. In comparison to bio-oil which is produced using pyrolysis, bio-crude has much less oxygen, however, the process of bio-crude production is still being developed (Figure 2.6).

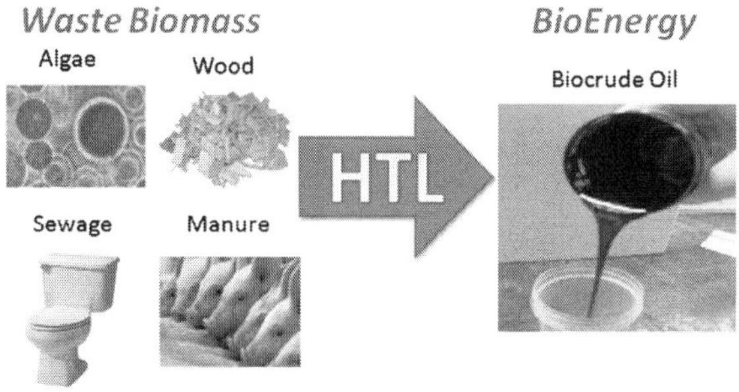

Figure 2.6: Biomass waste such as wood, manure, sewage, and algae can be put under hydrothermal liquefaction so as to produce bio-crude oil. This oil is still undergoing research as a substitute for petroleum.

Source: http://algae.illinois.edu/Projects/Hydrothermal.html.

Ethanol: Ethanol production through the fermentation of sugars is a classic method of conversion for corn, maize, and sugarcane on a large scale, especially in the United States, France, and Brazil. There are also some other countries, such as Zimbabwe, that have a functional fuel ethanol program with the main product being sugarcane. Programs in Europe and the United States normally convert excess food crops into byproducts that are useful. However, using corn and maize to produce ethanol is far from creating any competition with diesel and gasoline. In addition, these systems have an overall energy balance that is not at the very least favorable (Pimentel, 2008).

The PRO-ALCOOL program in Brazil is the main exception due to its high level of productivity of sugarcane. It is also possible to produce ethanol through the hydrolysis of lignocellulosic biomass. This option is highly efficient and potentially low-cost. In the United States and Sweden, hydrolysis techniques are gaining much more attention and are therefore undergoing development. If we lower barriers and combine the production of ethanol with efficient production of electricity from wood fractions that are unconverted such as lignin, the costs of ethanol would be very similar to those of gasoline currently, which is as low as \$0.13 a liter. The general conversion efficiency of the system would go up to approximately 70% (Pimentel, 2008).

Esters from Oilseeds: Rapeseeds and other oilseeds can undergo extraction and conversion to esters, and these have proven to be very efficient when it comes to replacing diesel. Rape Methyl Esters are produced at substantial quantities in the European Union and also in North America, but at a lesser extent. However, so as to compete with diesel, rape methyl esters require substantial subsidies. The energy balances for rape methyl ester systems are not as favorable as those that are used for perennial crops, meaning that per hectare, the net energy production is low. It is possible to improve these balances if the byproducts, such as straw, are also utilized as a source of energy.

Hydrocarbons, Hydrogen, and Methanol Through Gasification: Producing hydrogen and methanol with the use of traditional syngas conversion processes and gasification technology can help in providing a biomass conversion route that is highly attractive and which can be used as a long term option. Despite the fact that these concepts were earlier approved of, they were made unattractive by low oil prices. There is new technology, however, such as the production of liquid phase methanol and the technology for new gas separation, which has helped in offering higher overall conversion efficiencies and lower production costs. With the production of both electricity and fuel and large scale production, hydrogen, and methanol production with the use of lignocellulosic biomass can end up competing with diesel and gasoline. In addition, we can produce methanol and synthetic hydrocarbons from syngas with the use of Fischer-Tropsch synthesis.

2.2.1. Environmental Impacts of Biomass Energy Systems

When all biomass that is produced is utilized for energy, biomass energy can be carbon neutral. However, sustained production on the same piece of land for a long time period can bring about negative side effects on the fertility of the soil, the landscape, biodiversity, agrochemical use, and even water use. In addition, collecting, and transporting biomass leads to an increase in the use of infrastructure and vehicles, as well as emissions to the atmosphere (Figure 2.7).

Some of the impacts of biomass energy systems include:

- **Erosion:** This is an issue that is in relation to cultivating many annual crops in various regions. The crops that are best suited for energy production are perennials, having a land cover that is much better than that of food crops. With perennials, it is possible to keep the soil removal at a minimum during harvest. In addition, perennials form an extensive root system which adds to the soil's organic matter content. In general, it makes it possible to prevent diseases such as eelworms, thus improving the structure of the soil.

▲ Sheet erosion on the Springbok flats, Limpopo Province.

Figure 2.7: When biomass remains on the ground after land terrain, soil disturbance events and harvests, it causes soil loss. It is important to collect and recycle biomass for the production of energy so as to avoid this.

Source: https://www.grainsa.co.za/soil-erosion-in-south-africa---its-nature-and-distribution.

- **Water Use**: increasing the water use due to increased demand of vegetation that is new may end up becoming a problem, especially in semi-arid and arid areas. It is important to choose crops with water-use efficiency in mind. There are some crops that have extremely good water-use efficiency, such as some eucalyptus species, all in consideration of the quantity of water that is required for every ton of biomass produced. However, it is important to remember that planting eucalyptus plants over a large area can lead to an increase in the groundwater demand locally, thus affecting the level of available groundwater. However, improving the land cover can help in improving microclimatic conditions as well as water retention. This means that it is important to evaluate the impacts of the hydrological situation at the local level.

- **Agrochemicals:** the quality of surface water and groundwater is affected by pesticides, which in turn may affect animals and plants that are using this water. The effects that occur are all dependent on the chemical type, the application method, and the quantity used. Some of the perennial crops that have been used, including miscanthus, eucalyptus, poplar, and willow, have shown that the strict environmental standards have been met. In comparison to other food crops such as cereals, the agrochemicals rate of application per hectare are $1/5^{th}$ to $1/20^{th}$ for perennial energy crops.
- **Nutrients**: In agriculture, using manure and fertilizer in abundance has caused many different problems in the environment, especially in certain regions, such as unpotable water, eutrophication, a saturation of soils with phosphate, and nitrification of groundwater. The soil's heavy metal flux has also increased due to phosphates. However, in comparison to conventional agriculture, energy farming with perennial grasses and short rotation forestry needs much less fertilizer. When using perennials, we are able to obtain better recycled nutrients. In willow cultivation, the leaching of nitrogen can be ½ to a $1/10^{th}$ that that occurs with food crops, thus meeting the strict standards that are set for the protection of groundwater. Using plantation biomass leads to nutrient removal from the soil, all which need to be replenished. It is imp to recycle ashes so as to return phosphates and important trace elements to the soil, and this is a practice that is commonly used in Sweden and Austria.
- **Landscape and Biodiversity:** it is possible to criticize biomass plantations due to the fact that the biological species range that they support in much lower than that supported by natural forests. This information is however not always relevant. It would only be relevant if a biomass plantation replaced a virgin forest. However, if the establishment of plantations occurs on excess agricultural land or on degraded lands, these lands that have been restored have the ability to support an ecology that is much more diverse. There are many degraded lands, with estimates indicating approximately 2.1 billion hectares of available degraded land in countries that are developing. Such land surfaces need to be restored so as to improve certain aspects, such as climate control,

erosion prevention and water retention. A plantation design that is appropriate, inclusive of areas that have been put aside for animals and native plants that can fit in the landscape in a manner that is natural, can help in avoiding the issues that are generally in association with monocultures, with acknowledgment of the fact that a plantation of energy crops is not a confirmation of a monoculture (Figure 2.8).

Figure 2.8: The use of Large-scale biomass energy plantations cane either negatively or positively affect the biodiversity through: alteration of the habitat quality, soil erosion, pesticide use, quality of water, or other factors. These effects are partly dependent on the type of land use that the plantation is replacing.

Source: https://www.intechopen.com/books/advances-in-landscape-architecture/urban-landscape-design-and-biodiversity.

- **Other Risks Such as Disease and Fire***:* management of biomass production systems as well as landscaping can help in reducing the risk of disease and fire. This means that we need to pay more attention to these aspects in the future research, policies, and projects.
- **Conversion and End Use**: it is important to make sure that the processes for converting biomass to the intermediate energy carriers that are desired as well as their utilization for energy services need to meet environmental standards that are strict as well. Certain issues that may come up, such as emissions to the atmosphere, can be fixed in an easy manner with the use of technology that is both available and well understood. In Scandinavia, it is now common to have clean combustion of

biomass. Gasification makes it possible to clean fuel fast before it is combusted or processed further. Small conversion systems need to be handled with more care: there is available technology that allows them to meet the strict standards of emissions, however, this technology is expensive to purchase and operate, considering it is quite small.

2.3. WIND ENERGY

This type of energy source is broadly available, however, quite diffuse. Before the industrial revolution, wind energy was utilized widely as a power source, however, it was later displaced but the use of fossil fuels due to the differences in reliability and costs. There was an oil crisis in the 70s which led to an increased interest in wind energy technology for supplying power in remote areas, pumping water and grid connected production of electric.

There has been a lot of progress that has been made when it comes to the development of wind turbines to produce electricity. There has been production of more than 1.5 million water pumping wind turbines in many different developing countries that help in supplying livestock with water, especially in the remote areas. In Central Asia, Mongolia, and China, they are operating thousands of small battery charging wind generators (Nelson, 2013).

Onshore wind energy has a high technical potential of fulfilling energy needs—30,000 to 50,000 terawatt-hours every year. The wind energy economic potential is all dependent on the economics of the wind turbine systems and also of the different available options. In addition to the costs for the investment, the other main aspect that comes into play during the determination of the wind turbine system economics in the yearly energy output, which in turn undergoes determination through certain parameters, including roughness of the surrounding terrain, turbulence intensities, statistical wind speed distribution, and the average wind speed. Due to the sensitivity to the speed of wind, it is not possible to determine a straightforward potential of wind energy at one certain site. As of now, there is production and publishing of wind energy handbooks and maps, as well as meteorological measurements, making it possible for wind project developers to assess the economic performances of their projects in the long term, in a much easier manner (Nelson, 2013) (Figure 2.9).

Figure 2.9: It is very rare to find that wind energy has caused air pollution, unlike other power plants that are reliant on combusting fossil fuels such as natural gas or fuels, all which emit sulfur dioxide, nitrogen oxide and particulate matter. In addition, turbines do not produce emissions that lead to greenhouse gases, smog, or acid rain.

Source: https://www.energy.gov/eere/wind/advantages-and-challenges-wind-energy.

In countries that are densely populated, the most appropriate sites for this technology are inhabited, and resistance from the public makes it much harder to realize new projects at costs that are acceptable. This is actually the main reason why the Netherlands and Denmark are developing projects offshore, despite having economics that are less favorable. The United Kingdom and Sweden are developing projects offshore so as to ensure preservation of the landscape (Sayigh, 2012).

Resources that are found offshore are actually much bigger than those found onshore, however, for them to work as efficiently as possible, it is important for them to be close to electric infrastructure.

Some of the technology developments include:

The Wind Turbines Become Bigger: From the first time that wind energy technology was modernly utilized in the mid-70s, the unit size of commercial machines has grown gradually. In the mid-70s, the wind turbine had a generating capacity size of 30 kilowatts, with the diameter of the rotor being 10 meters. In 1998, the largest units to be installed had a generating capacity size of 1,640 kilowatts, with the diameter of the rotor being 66 meters. By the time it was reaching the late 90s, there were global installations of 460 units each having a generating capacity size of one megawatt or more. As of now in the market, we are having the introduction of turbines that have a generating capacity size of two megawatts, with the diameter of the rotor being 70 meters. In addition, engineers are working on developing machines that have a generating capacity of 3 to 5 megawatts.

The Trend is Driven Towards Larger Machines due to Market Demands*:* There will soon be development of expectations that offshore potential, less visual impacts on the landscape per unit of power that has been installed, and economies of scale. There is an expected installation of wind turbines with 1,500 kilowatts. It is however important to note that the optimum turbine size when it comes to public acceptance, impact, and cost, is different for offshore and onshore (both remote and nearby) applications.

Wind Turbines Become More Grid-Compatible and Controllable: Apart from switching the machine off and on, the output of wind turbines that are stall regulated is hardly controllable. There are variations in the output due to the speed of the wind until it reaches the rated wind speed value. With bigger turbines, the use of aerodynamic stall phenomena to machines that are structurally compliant has become much more difficult, and because of this, there is application of blade pitch control systems. For structural reliability and dynamics, there should be combination of a blade pitch system with a variable speed electric conversion system. Electronic components that are currently being used have made it possible for designers to have full control of the output, within the range of wind speed that is operational, and ensure production of incredible power quality. These are developments that have made is possible to integrate wind turbines with the electricity infrastructure and in the end provide a higher level of penetration. This is a benefit that mainly interests weak grids, mainly in remote and rural areas that have a lot of wind (Sayigh, 2012).

There Will Be Fewer Components in Wind Turbines: Designers are now looking for technology that has less components so as to achieve greater maintainability and reliability, and lower costs, such as generators that run slow and are directly driven, with passive blade pitch control and passive yaw (Figure 2.10).

Figure 2.10: The turbines can either rotate in a vertical or horizontal axis. They may also either have no blades or include blades.

Source: https://www.cnbc.com/amp/2020/01/17/work-starts-on-huge-wind-farm-that-could-power-4point5-million-homes.html.

Special Offshore Designs are Being Developed: Having the initial offshore wind farms in Europe, there has been development of dedicated turbine technologies in the open sea for wind farms that are large. It is possible to quickly correct outages onshore so as to only lose a small amount of energy. However, offshore, there is a limited window for repairing or replacing components. Complete installations have a high cost and this implies that large wind turbines are being used, and these may have generating capacities of 3 to 6 megawatts. Some of the design features offshore include: integration with external conditions (both wave loading and wind), corrosion protection, transport, and electricity conversion systems, and novel installation concepts.

The Project Preparation Time is Now Longer than the Time to Market: Despite the fact that in some countries there is a temporary supply shortage for wind turbines, there is high competition among manufacturers. One of the main ways to remain at the top of the competition is by ensuring the implementation of component improvements and innovations so as to lower costs. There is also reduction in the time to market new products (2–3 years). Due to this, it is at the commencement of the wind farm construction that the technology being used becomes outdate.

2.3.1. System Aspects of Wind Technology

There are also some aspects that need to be understood about the wind technology system. These include:

Wind Turbines are Able to Deliver Energy, However Provide Little Capacity: Due to the fact that wind energy is intermittent, the delivery of energy by the wind turbines occurs, however at a small capacity value—mostly <20% of the wind power that is installed. In addition, there is a drop in this percentage with increase in the wind turbine's penetration, needing much more backup power for a supply of energy that is reliable. Despite this, it is possible to transform wind-generated electricity from intermittent to base-load power, especially if it has undergone combination with, for example, compressed air energy storage. This makes it possible to achieve a high capacity factor with a small economic penalty.

The Predictability of Wind Power Increases: Meteorologists have provided research work that has proven the ability to predict a wind farm's output in advance with the use of computer programs that are able to optimize the fuel and operational costs of electricity production parks that are found in the region. This helps in increasing the wind power's capacity value as well as the value of the produced electricity.

It is Possible to Adjust the Capacity Factors: It is possible to have wrong conclusions and decisions due to some general misconceptions. The annual energy output is referred to as the capacity factor, and this is dependent on the wind turbines and local winds. By optimizing the characteristics of the turbine to the local wind regime, it is possible to optimize the capacity factor (which is normally 20 to 25%), without the loss of excessive energy output. However, having capacity factors that are extreme (40%), means that there will be an automatic loss of potential energy output.

Renewed Interests in Systems that are Autonomous: In the mid-1980s, the use of wind turbines in areas that are isolated and lack energy infrastructure grew in interest. It is possible to distinguish two systems:

- **Hybrid Systems:** in these systems wind turbines operate in parallel with other technologies, for example, a diesel set so as to ensure that fuel consumption is reduced, in addition to reducing repairs and maintenance; or a diesel generator that has been out together with a battery storage unit.
- **Standalone Units:** these are used for cooling and desalination, watering cattle, domestic use, pumping water for irrigation, or for charging batteries.

There has been development and testing of more than 30 hybrid systems that are experimental, most of them ending before being followed up commercially, due to components that are expensive and unreliable. There is however a revival of the interest in standalone and hybrid systems, due to the search for new renewable energy systems markets. In addition, this has been influenced by the cost and performance improvements of power electronics and wind turbines. For the systems to successfully enter the market, they have to be modular, and there has to be the introduction of standards for subsystems and components (Sioshansi, 2011) (Figure 2.11).

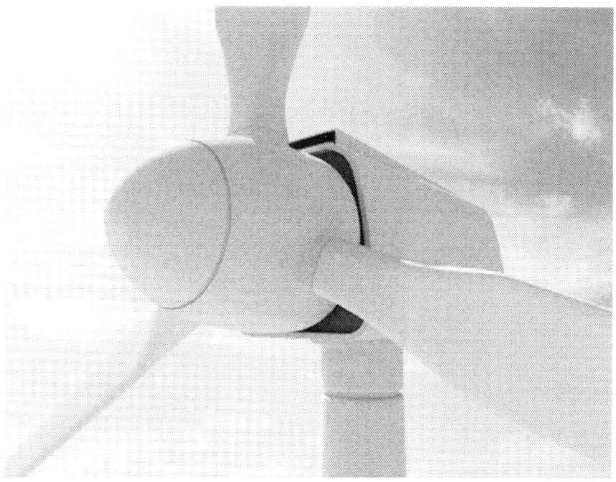

Figure 2.11: The rotor blades on wind turbines are being lengthened so as to raise the level of efficiency. This means that they need to be resistant to fatigue, light, strong, and stiff. This can be achieved with the use of composites such as epoxy and polyester. Carbon fiber and glass fiber can be used as reinforcements.

Source: https://scitechdaily.com/successful-test-for-superconducting-3-6mw-wind-turbine/amp/.

2.3.2. Environmental Impacts of Wind Technology

We will use the three main phases of a wind turbine project to illustrate the environmental aspects. These include: construction and manufacturing, the turbine's lifetime normal operation, and decommissioning.

Construction and Manufacturing: During the construction civil works or of a wind turbine, there are no manufacturing processes or exotic materials required. Under typical Danish conditions, a large wind turbine's energy payback time is approximately three to four months.

Normal operation: Some of the negative impacts on the environment in connection with wind turbine use are: electromagnetic interference with radar signals, television, and radio; moving shadows created by the rotor; impact on bird life; visual impact on the landscape; and acoustic noise emission. Generally, most problems are caused by the visual and noise impact. Designers are not able to increase the tip speed of the rotor blades due to acoustic noise emission, thus preventing them from increasing the drive train shaft's rotational speed, which would help in reducing the cost of generators or gearboxes. Research on aero-acoustics has helped in designing blade and tools configurations that help in making blades much quieter, thus leading to a reduction in the distance required between houses and wind turbines. If turbines are properly located, the impact on birdlife becomes a minor issue. When operating normally, wind turbines do not release any emissions to the atmosphere. This means that the main way to ensure reduction of carbon dioxide emissions is dependent on the fuel mix pf the fossil fuel plants that are being used to power the wind turbine.

Decommissioning: Due to the fact that all components are conventional, the methods used to recycle to decommission the wind turbine are also conventional. Majority of the blades are produced using carbon fiber reinforced plastics or glass, both undergoing processing through incineration. So as to make a replacement of carbon and glass so as to end the material use cycle, there is now application of wood composites and development of bio-fibers.

2.4. HYDROELECTRICITY

Over the past years, people have viewed hydroelectricity as a source of energy that is renewable, more economically attractive, non-polluting, and non-exhaustible than other options. Despite the fact that it is possible to build a finite number of hydroelectric plants, only 1/3rd of the plants considered as economically acceptable are tapped. In comparison to thermal plants, hydropower plants release much less greenhouse emissions. Those that are released are due to vegetation decay in areas that are flooded, and also by the excessive use of cement in constructing dams (Spellman, 2014).

Sadly, hydropower plants cause some negative local impacts because they use both ecological and social rivers, and these issues are becoming much more important as people learn everyday how their living standards are affected by these impacts. Most of the technology utilized for the generation of hydroelectricity are capital intensive, just like other renewable sources; however, in comparison to nuclear and thermal options, they have less operational costs. The major barrier for the growth of hydroelectricity in countries that are developing is the high initial cost, yet this is where most of the economic potential that is untapped is (Spellman, 2014) (Figure 2.12).

Figure 2.12: Hydroelectric power plants are found in dams that have impounded rivers, leading to a raise in the amount of the water behind the dam, leading to the creation of a head that is as high as possible.

Source: https://www.azom.com/amp/article.aspx?ArticleID=18262.

2.4.1. Hydroelectric Technology Development

Technologies to reduce power generation costs and dam construction: Technology used for the generation of hydroelectricity is normally referred to as mature technology, which means that the chances of it advancing are low. This is the case for the cost and efficiency of conventional turbines, due to the construction of a large number of units leading to a design that is already optimized. However, for small scale hydropower, it is possible to achieve more technical development. For example, further development of innovative turbines, new materials, submersible turbo-generators, electronic telemetry and control, induction generators, and the use of variable speed turbines at low heads (Sayigh, 2012).

On the construction of dams, further progress has recently been made, mostly with roller compacted concrete dams. These have a lower cement content and the fact that the placing of the concrete is mechanized allows for one to yield unit costs that are relatively low, of around 30 to 40 dollars per cubic meter of dam body, which is approximately lower than half the cost of concrete that is conventionally placed. Because of the rapid placement of concrete with the roller compacted concrete dams, it is possible for the dam to grow daily by approximately 60 centimeters, meaning it is possible to construct a dam that is 200 meters high in less than one year. With the roller compacted concrete dams, during construction, the river diversion is normally in-river, instead of using diversion tunnels, which helps in saving money and time. Many dams that previously appeared economically steep have been made feasible by the roller compacted concrete technology. So as to achieve structures that are smaller, dams with inflatable rubber weirs that go up to 15 meters high, and geo-membrane lining that can go up to 80 meters' high are slowly being accepted as alternatives to earth-fill/rock-fill dams and concrete weirs.

There is now a better understanding of other parts of the operational system, such as spillways, making it possible to use spillway chute discharges per meter width that are highly specific, leading to cost saving. There is also an increasing attraction to tunnel-boring machines. The reason to this is because they do not disturb the landscape. Control rooms and powerhouses are being designed in a manner that will help in cutting manufacturing time and costs of hydroelectric equipment. It is expected that creating advancements in technology can help in reducing costs, however, it must be done in small quantities at a time because the current technology is already well-optimized (Sayigh, 2012).

The unit cost of large scale hydropower is much lower than of the small scale hydropower. However, choosing sites that are favorable, refurbishing old sites, avoiding cofferdam use during the installment process and using existing civil works for the purpose of controlling floods as well as the administrative structures that already exist helps in reducing the costs for electricity production.

Technology for the Reduction of Ecological and Social Impacts: There is a lot of critics surrounding the production of hydroelectric power, especially when building larger dams, and due to this, modern construction aims at including certain technologies in the system design that help in reducing the negative ecological and social impacts. This is discussed in more detail in the next section (Figure 2.13).

Figure 2.13: There is a wide variation of the reservoir size that is created by a hydropower project, mostly dependent on the topography of the land and the size of the hydroelectric generator.

Source: https://www.kidsdiscover.com/teacherresources/whats-good-and-whats-bad-about-hydropower/.

2.4.2. Environmental and Social Impacts

There is a significant amount of land that has been flooded due to the generation of hydroelectricity. If the average energy density of mew plants is maintained, it can lead to the flooding of an extra 60 million acres of land so as to avail two-thirds of the economic potential. The figure provided may not look high in relation to the amount of land that is needed for the production of biomass energy, however, areas surrounding the river are the highly inhabited especially in rural areas (Sioshansi, 2011).

It is possible to minimize the different environmental impacts in areas that are flooded by conveniently choosing sites that make it possible to ensure storage of large quantities of water in areas that are small, such as canyons. The overall cost of hydropower production is increasing with the increased responsibility to ensure preservation of the environment. As hydroelectricity enters the market, it is hard to come up with a proper solution to deal with marginal cost pricing. It is also possible for operators that have high costs to find themselves in environments that are restructures with insufficient and old generating plants. In the United States, there is decommissioning of certain dams that are associated with the production of power, because they impose a significant onus for biodiversity or disturb commercial fishing.

One of the impacts that is well understood is the displacement of inhabitants from the area that is flooded, and mitigation of this issue can lead to the use of high costs. It is always expected to have strong criticism for hydroelectric projects that require to relocate many people. The most important aspect is to build trust with the people who are being affected by the displacement and having resettling, the authorities, and the developer; these people need to feel and know that they are being taken seriously and they matter (Sioshansi, 2011).

Some of the major issues that arise when constructing these hydroelectricity dams are: downstream impacts, deterioration of human health, water quality standards, biodiversity perturbation, sedimentation, and fish biodiversity and fish amount. These are discussed below:

- Changes in fish biodiversity and fish amount-there is a need to develop technologies that are able to ensure preservation of subsistence and the production of commercial fish, as well as the biodiversity of fish. So as to achieve a quantitative understanding of how the fish respond to many different stresses within a turbine, as well as for the development of biological performance criteria so as to utilize in designing advanced turbines, there is recommendation for further research and development. Including passage facilities, for example fish ladders, is becoming necessary for the renewal of contracts required for operating dams in the United State. Where such technology is not helpful, such as in tropical countries, one of the solutions could be luring fish into elevators or containers using electricity. Due to the fact that majority of the newer dams are constructed in tropical countries, it is important to ensure that extensive studies are carried out for the identification of rare and new species as well as determining if these species can survive in adjacent rivers that are not slated for damming.
- Sedimentation-when catchments are developed, there is a strong increase in sedimentation. If a landslide occurs, or a heavy flood, the reservoir would be filled up sporadically with a large amount of land. It is possible to minimize these problems through using reservoir density currents, sediment dredging, sediment sluicing, sediment flushing, sediment bypassing, debris dams, and watershed management.
- Biodiversity perturbation-conserving the demands of diversity is important, especially when it comes to making sure there is no net loss of species. This means that there is a need to acquire knowledge on the already available components before building

a dam, and this has proven to be difficult. The main measures that need to be taken to ensure conservation are selection of the reservoir size and site selection. In practice, conserving onsite biodiversity is mainly dependent on making sure that large areas do not get flooded, especially habitats that are intact, such as tropical forests, as well as on the conservation of a perpetuity offset.

- Quality of water-generally, the disturbance of water quality is mainly due to large amounts of biomass that have been left in areas that are flooded as well as through reservoir filling. It is possible to mitigate this through filling up the reservoir at an acceptable rate, or by removal of the biomass. After the reservoir is filled up, it is possible for thermal stratification to occur having a water depth greater than 10 meters as well as a long water residence time. Water of warmer or colder temperatures that that released by the river without a dam, can be released through reservoir stratification, and this could lead to both negative or positive impacts on the fishery in the river. It is possible to minimize this through: aeration and positive mixing by compressed air or fountain jets; multilevel outlet works for mitigating the effects of downstream; in-reservoir de-stratification; and changes in the configuration of the inlet structure. However, there is not enough knowledge to use this technology and further research on development needed (Figure 2.14).

Figure 2.14: Above is an example of a downstream impact of hydroelectricity plant dam: disrupted flow of water and sediments. Others include: decreased fish populations, lower crop production, poor water quality, and reduced biodiversity.

Source: https://www.deccanherald.com/state/water-released-linganamakki-687518.html.

- Deterioration of human health-epidemics can be caused by reservoirs, specifically three diseases that are related to water: Japanese B encephalitis, schistosomiasis, and malaria. It is possible to avoid the proliferation of encephalitis and malaria with chemotherapy and chemicals. However, Plasmodium protozoan parasites and mosquitoes have started to gain resistance making malaria much harder to deal with. It is better to control schistosomiasis with chemotherapy.
- Downstream impacts-upstream resettlement upheavals can be exceeded by downstream social impacts, and they need much more attention that they are currently being given. Moisture deposition and cessation of annual fertile slit can lead to a decline in yields, an impairment in grazing, a decline in wildlife and fish, and erosion at the river's mouth, all because of a decline in suspended particles that help in replacing land that the ocean has washed out. Additionally, a decrease in the agricultural yields and availability of water leads to an increase in water competition as well as other scarce resources. Also, constructing dams forces people who are used to cyclical floods to suddenly switch and adapt to livelihoods that are rain-fed. It is possible to mitigate some of these issues through various level off-takes so as to promote the temperature flexibility of the water in accordance with the needs of the downstream, as well as others through measures that help in reducing the level of reservoir stratification, inclusive of shorter water residence time and local mixing.

2.5. SOLAR ENERGY

Figure 2.15: Solar panels are the most common technology for collecting sunlight and directly turning it into electricity. These are used all over the world as

an alternative for other expensive and environmentally unfriendly sources of electricity.

Source: https://oilprice.com/Energy/Energy-General/Low-Solar-Panel-Prices-Spark-Surge-In-Adoption.amp.html.

This involves converting energy from sunlight to electricity, and this can be done in a direct manner with the use of photovoltaic's (PVs), or in an indirect manner with the use of concentrated solar power. Concentrated solar power systems normally utilize mirrors or lenses as well as systems for solar tracking so as to focus a wide sunlight area into a smaller beam. PV cells normally work by converting light into an electrical current with the use of the PV effect (Kalogirou, 2009) (Figure 2.15).

Initially, PV's were only utilized as an electricity source for medium and small-sized uses, from remote homes that use an off grid rooftop PV system for power to a calculator that is uses a single solar cell for power. The development of the first commercial concentrated solar power plants was in the 1980s. There has been an increase in the number of grid connected solar PV systems with a decrease in the cost of solar electricity. Solar PV is increasingly becoming a low cost, low carbon technology for harnessing renewable energy from the sun. As of now, the largest PV power station globally is in India. It is called the Pavagada Solar Park, and has the ability to generate up to 2050 megawatts (Kalogirou, 2009).

2.5.1. Solar Energy Technologies

There are many industrialized countries that achieved the installation of a significant capacity of solar power into the electricity grids for the supplementation or provision of an alternative to the conventional sources of energy, while many other developing countries have chosen to completely use solar to reduce the level of dependence on imported fuels which are quite expensive. Remote resources for renewable energy are able to displace the consumption of fossil fuels through long distance transmission. There are two main technologies that solar power plants utilize:

- PV systems: these utilize solar panels which can be found in ground-mounted solar farms, or on rooftops, and these directly convert sunlight into electricity.
- Concentrated solar power: this is also referred to as concentrated solar thermal. These plants normally make steam through solar

thermal energy, and this steam is thereafter converted into electrical currents with the help of a turbine.

2.5.1.1. Photovoltaic Cells

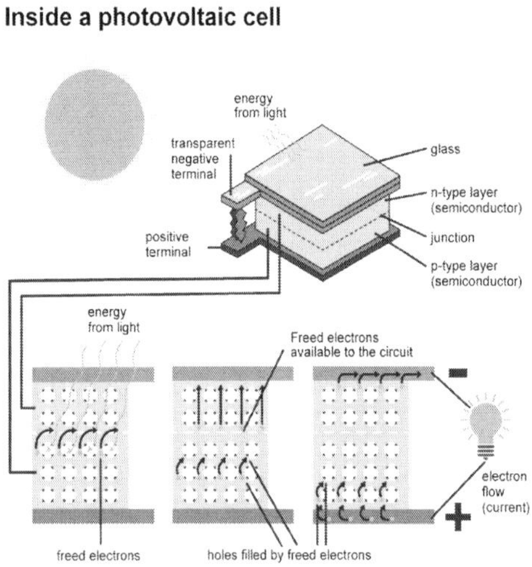

Figure 2.16: Above is an image showing the inside of a photovoltaic cell. This shows how the cell is able to use sunlight to produce electricity.

Source: https://www.eia.gov/energyexplained/solar/photovoltaics-and-electricity.php.

Photovoltaic System Components: There is need to build a complete system so as to make use of electricity from PV cells. This is inclusive of electricity storage, support structures, and the comprising electronic parts (Figure 2.16).

Size and type of PV systems: These can be utilized for many different applications, from grid-connected large power stations and rooftop systems, to small standalone units and consumer products for rural use.

Need for Storage: Due to the fact that PV modules provide a source of energy that is intermittent, majority of the standalone systems have battery storage for the provision of energy at night or days that have a low amount of sunshine. These batteries are usually lead-acid batteries. In some situations, it is possible for batteries to retain energy for longer time periods. When they use PV systems that are grid-connected, the grid functions as a virtual storage, meaning that PV's feed electricity into the grid therefore reducing the use of fuel by gas, oil or coal-fired power plants.

Performance ratio of Photovoltaic Systems: It is necessary to make predictions of the amount of energy that a specific PV system feeds into a grid. However, for this to be effective, there is a need for reliable information on the insolation in the system's plane, the system losses, and the system power under standard test conditions.

2.5.1.2. Concentrated Solar Power

There are many different concentrating technologies that are available: the solar power tower, the dish Stirling, the compact linear Fresnel reflector, and the parabolic trough. There are also different methods that are used for tracking the sun and focusing the light. All these systems have a similar mechanism. They all heat a working fluid through the concentrated sunlight, and this is used for the generation of power and the storage of energy. An approximate 24-hour of electricity generation is possible with efficient thermal storage.

Parabolic troughs have reflectors that are linear which help in concentrating light onto a receiver that is found along the focal line of the reflector. There is a tube positioned along the linear parabolic mirror's focal point and this is the receiver in the system. It is filled with a working fluid. This reflector specifically tracks the sun along a single axis, thus follows it during daylight hours. Of all the solar technologies, parabolic trough systems provide the best land-use factor (Sioshansi, 2011).

The compact linear Fresnel reflectors utilize multiple thin strips of mirrors, instead of parabolic mirrors, for concentrating sunlight onto working fluid-filled tubes (two of them). Thus provides the benefit of using flat mirrors, and these are much more affordable in comparison to parabolic mirrors. Another advantage is that it is possible to place more reflectors in the same amount of space, thus making it possible to optimize the available sunlight. This technology can be used in either more compact or large plants.

The Stirling solar dish is generally a combination of a Stirling engine, which is normally used for driving an electric generator, and a parabolic concentrating dish. This technology has an advantage over the PV cells: they provide a higher level of efficiency when it comes to the conversion of sunlight into electricity. They also have a longer lifetime. Among CSP technologies, parabolic dish systems provide a higher level of efficiency (Sioshansi, 2011).

Solar power towers normally utilize several tracking reflectors which are generally referred to as heliostats for the concentration of light on a receiver that has centrally been placed at the top of a tower. These towers are able to achieve higher levels of efficiency when it comes to converting thermal energy to electricity, in comparison to linear tracking CSP schemes, as well as better capabilities for storing energy in comparison to the dish Stirling technologies (Kalogirou, 2009).

2.5.2. Environmental and Social Impacts of Solar Energy

The sun provides an incredible resource for producing clean and viable electricity without toxic contamination or global warming discharges. The prospective environmental impacts related with solar power—including land use, habitat loss, water utilization, and hazardous materials utilization in manufacturing—can vary immensely based on the technology, which comprises of two broad categories that are: PV solar cells and concentrated solar thermal plants or (CSP).

The scope of the system—varying from small, distributed rooftop PV sequences to massive utility-grade PV and CSP projects—likewise plays a big role in environmental impact levels. Based on their location, bigger utility-grade solar facilities may address concerns around land degradation and loss of habitat caused by other energy sources like geothermal. Total land area requirements may vary based on the technology, the site topography, and intensity of solar resource. Approximations for utility-grade PV systems vary from 3.5 up to 10 acres per megawatt, whereas estimates for CSP amenities are around 4 to 16.5 acres for each megawatt.

However, unlike wind infrastructure, there is less prospect for solar energy investors to share land resources with agricultural uses. Nevertheless, adverse land impacts from utility-grade solar systems can be reduced by putting them at low-scale locations like brownfields, abandoned mine lands, or existent transport and transmission corridors. Likewise, smaller grade solar PV arrays, which may be built on homes or commercial buildings, have lesser land-use impacts.

2.6. GEOTHERMAL ENERGY

For thousands of years, this is type of energy has been utilized for washing and bathing, however, in the 20th century, it was developed for use on a large scale for the production of electricity, industrial energy use, and space heating.

Iceland was the first to develop a large municipal district heating device in the 1930s. We have commercially used geothermal energy for approximately 70 years, and around 40 years for both direct use and generation of electricity. There has been an increase in its use over the last three decades. Around 80 countries have geothermal resources, and approximately 50 of them are actually using geothermal energy (Figure 2.17).

Figure 2.17: The geothermal resources of earth are more than enough for supplying the energy needs for each one of us on a global scale, however, only a very small part of it can be exploited in a profitable manner, because exploration and drilling for these resources is costly.

Source: https://climatechange-theneweconomy.com/12291-2/.

There are many different gases found in geothermal fluids, with Carbon dioxide and Hydrogen at the top of the list, and it may also contain some hydrogen sulfide and even much smaller quantities of boron, radon, mercury, and ammonia. The amounts of these gases are all dependent on the field's geological conditions.

There is concentration of majority of the chemicals in the disposal water, which is commonly injected back into the drill holes, and therefore it is not released to the environment. The gases are not released in harmful concentrations, and removing them, such as hydrogen sulfide alone from the steam produced by the geothermal energy, is a common matter in geothermal plants that have high gas levels (Sayigh, 2012).

High-temperature geothermal fields that are utilized for the production of electricity have a wide range of carbon dioxide emissions for the global production of electricity, being 13 to 380 grams a kilowatt-hour, which is much lower than fossil fuel power stations. There are also lower levels of sulfur emissions for geothermal than there are for fossil fuel plants.

Low-temperature geothermal resources produce gas emissions that are only a fraction of those that are produced by high temperatures fields that are utilized for the production of electricity. Low-temperature waters tend to produce a very small gas content— in Iceland, the amount of carbon dioxide in such waters is much lower than that of cold groundwater. In basins that are sedimentary, for example, the Paris basin, the amount of gas may be too much to release to the atmosphere. In such situations, a closed circuit is used to make sure that the geothermal fluid is kept at pressure and the degassed fluid is injected back into the reservoir. When there are conventional geothermal schemes in basins that are sedimentary, there is a tendency for the production brines that have to be injected back into the reservoir and this means that they can never be released to the atmosphere. This means that the carbon dioxide emission from these technologies is zero (Sayigh, 2012).

CHAPTER 3

Water Purification Technologies

CONTENTS

3.1. INTRODUCTION

In a bid to improve the environment for the human race, several efforts are implemented to mitigate negative impacts. They include purification of water, making it safe for human consumption. As water is an essential element it requires major efforts to be put in place to make this basic need available to the population. Water purification is the process in which undesirable chemicals, gases, and suspended solids are removed from the water. Water purification is essential. The importance may be attributed to the presence of water pollution to some greater extent. It is also important because in the urban areas, water is supplied to the homes and most of the water pipes supplying the water flow from treatment plants to the homes. Some of these pipes uses have been in place for longer periods of time and maybe worn out. This means that there is a high probability of the water being contaminated. It may reach to some extent whereby the water is saturated with contaminants. The contaminants are a threat to the wellbeing of individuals as it may undermine the health of individuals (Cheremisinoff, 2002).

The contaminated water may introduce harmful bacteria to the body. The bacteria may infest the body and damage the immune system. Apart from that the contaminants could lead to damaging of the plumbing and household appliances. There are a variety of contaminants that could be present in water. They include aluminum, arsenic, barium, cadmium, chloride, chromium, copper, fluoride, lead, nitrates, mercury, and silver. There are different types of water purification methods. They are used according to their availability. There are various reasons why water purification is conducted. Apart from making water safe for drinking, there is also to allow medical, pharmacological chemicals, and industrial applications. There is a classification done on the water purification process (Figure 3.1).

- The physical process which involves filtration, sedimentation, and distillation;
- The biological process which involves the use of either slow sand filters or biologically active carbon;
- The chemical process which involves the use of chlorination and flocculation; and
- The use of electromagnetic radiation which involves the use of ultraviolet (UV) light.

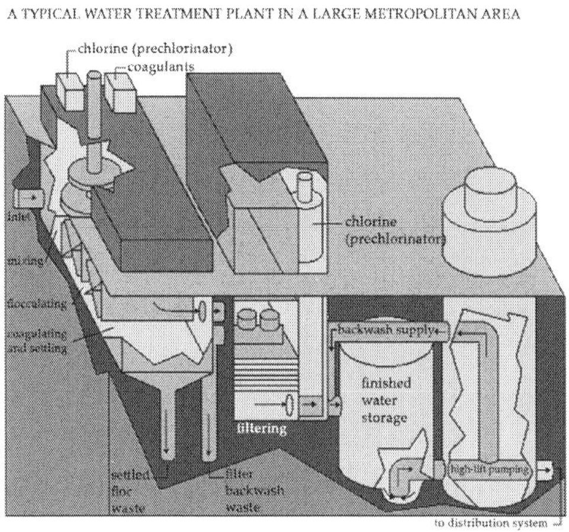

Figure 3.1: The water purification processes aim at minimizing the concentration of certain particulate matter. Examples of particulate matter are parasites, bacteria, suspended particles, algae, bacteria, fungi, and viruses. They also minimize the concentration of a range of dissolved particles.

Source: https://www.britannica.com/topic/water-purification.

Water purification allows the water quality standards set by the government to be met. The water quality standards are a guideline that indicates the maximum and minimum concentration of contaminants that could be present in water. This is dependent on the use intended for the water. This will involve the conducting of analysis to determine whether they provide water that has met these standards. The inspection however will not employ the use of visual inspection. This is because visual inspection will not facilitate the identification of certain micro-organisms as they are microscopic. Some of the techniques used in water purification such as boiling are not suitable for treating water containing certain contaminants (Johnson, 2006).

For instance, water containing many metals as well as chemicals may not be treated entirely through the use of boiling. The use of activated carbon filters may not be entirely effective. Initially, some of the water sources produced water that was said to be safe for drinking. However, with the rise in water pollution, the water from these sources need to be tested before the public is allowed to consume it. The analysis of water may employ the use of both chemical and microbiological methods. This is important as it allows one to select the most appropriate method that can be used in the purification of water. Recent research conducted by the WHO indicates that a large percentage of the population is unable to access water safe for drinking.

This is known to contribute to the large numbers recorded of people suffering from water-borne diseases such as cholera. Many individuals have succumbed to diarrhea. This has contributed to the pressure mounted on the government to provide safe water. However, the provision of safe water for drinking may not be entirely dependent on the government. The citizens themselves can contribute by making use of water purification technologies that can be done at home. These technologies involve the use of chlorination, filters, proper storage, and also water disinfection. This has been taken up by most of the family homes steads. The impact has been the reduction of the number of death that can be attributed to waterborne diseases (Cheremisinoff, 2002).

3.2. IMPORTANCE OF WATER PURIFICATION

Apart from water purification contributing to the removal of unwanted sediments in water, there are other reasons that contribute to the importance of conducting water purification. They include the following:

Reduction in Chlorine Levels: There are significant amounts of chlorine present in water. If they reach certain levels, then they pose a threat to human existence. For instance, in the high levels, the chlorine may be carcinogenic causing a risk of cancer. It may also lead to problems such as cardiovascular problems and asthma among others. There are techniques that can be used to remove the chlorine from water. They may involve the use of a water plant purifier. It makes use of industrial water purifier thereby reducing health problems.

Cost-Effective: Instead of constantly purchasing clean water, one may end up choosing to purchase a water purification system which in the long run will prove to be cost-effective.

Enhanced flavor: In most cases, the impure water tends to have a metallic taste among other unpleasant tastes. By purifying the water, germs will be removed from the water and it will improve its flavor.

Convenience: The use of water purification system assists in removing contaminants and in the process it is less-time consuming. This is convenient in the case where there is the need for large amounts of water. It is therefore an easy way to get rid of germs and saves time.

Reduced Bacteria Levels: Most of the impure water is said to be composed of bacteria. Though there are some bacteria that is useful to the body, there are some that are harmful. If the body contains large amounts of bacteria, it may be beyond the balance required by the body and thereby dangerous. The water purification system seeks to remove life-threatening bacteria from water. This minimizes the risk of the body being exposed to such types of bacteria.

Reduce risks of Medical Conditions: As dangerous contaminants are removed from water; it minimizes the risk of dangerous diseases such as cancer. This is attributed to the fact that impurities such as bacteria and chlorine are removed. This reduces potential health risks.

Skin Problems: Impure water contains contaminants such as chlorine and other contaminants which may contribute to the skin being irritated. Purified water reduces the chances of one being exposed to this potential threat.

Stronger immunity: Pure water is said to contribute to a healthy immunity. The reverse is also true. Impure water causes a potential health risk to the immunity more so to children because it has the ability to completely destroy it. Therefore, good quality water contributes to a healthy state of the immune system.

Reducing the Chances of Having Birth Defects in New Born Babies: For pregnant mothers, what they consume is important as it may contribute to the state of health of their unborn babies. Contaminated water may contain chlorine which is among the contaminants that can lead to a mother giving birth to a child with birth defects. Some of the birth defects include spina bifida among others. It is therefore important for the mother to consume pure water.

Well Working Digestive System: It has been known that most of the problems related to digestion has been as a result of people consuming water having contaminants. It is advised that people should take pure water as it leads to the improvement of metabolism. The good metabolism will contribute to a good digestive system. Pure water also lacks toxins. This means that the body be detoxified.

It removes Gastrointestinal Problems: The digestive system should be able to digest heavy meals. This is made possible by incorporating water into the diet. Therefore, it is important that the water used in this process be pure and healthy. This assists in preventing the gastrointestinal diseases.

There are various processes involved in water purification plants as there are a variety of treatment options used.

3.3. WATER TREATMENT OPTIONS

Some of the treatments options include:

Water Fluoridation: This option involves the addition of fluoride in water. The fluoride is incorporated after the disinfection process is done. In some countries, the fluoridation is made possible through the use of hexaflouric acid. Once introduced in water, it undergoes decomposition that results in the yielding of fluoride ions.

Water Conditioning: This is done with the aim of minimizing the effects of hard water. This process involves water being subjected to high temperatures which lead to the hard salts being heated. They are afterward deposited as decompositions of bicarbonate ions that are later on precipitated out of the solution. The precipitation is achieved through the use of soda ash. The precipitate in most cases is calcium carbonate which can be used in the making of toothpaste.

Plumb Solvency Reduction: This option of water treatment is used in areas that have naturally acidic water that have relatively low conductivity. This implies that the water has the capability of dissolving lead. The treatment involves the addition of relatively small amounts of phosphate ions. It is followed by an increase in pH. This causes a reduction in the plumbo-solvency as it causes the creation of insoluble lead salts.

Radium Removal: This treatment option is used in places where the water sources have radium. This is a radioactive chemical element. This treatment involves the use of ion exchange to remove the element. The water then undergoes conditioning. The wastes from the water are usually low-level wastes.

Fluoride Removal: For some water sources they have high levels of fluoride which may be harmful to individuals. The treatment involves the minimizing of fluoride levels. This is made possible by the use of activated alumina as well as bone char filter media.

These are among some of the treatments used in relation to water purification. These treatments make use of some of the water purification technologies. The technology is specific to a certain contaminant which it effectively removes. Some of the water purification technologies are listed below.

3.4. TECHNOLOGIES USED IN WATER PURIFICATION

Some of the technologies used include:

3.4.1. Carbon Filtration

This method makes use of carbon filters. This method works by allowing water to flow through an activated carbon filter cartridge. The carbon then absorbs the objectionable odors, organic compounds, flavors, and other dangerous chemicals. The technique is called carbon filtration as it employs the use of a carbon filter cartridge during the filtration process (Figure 3.2).

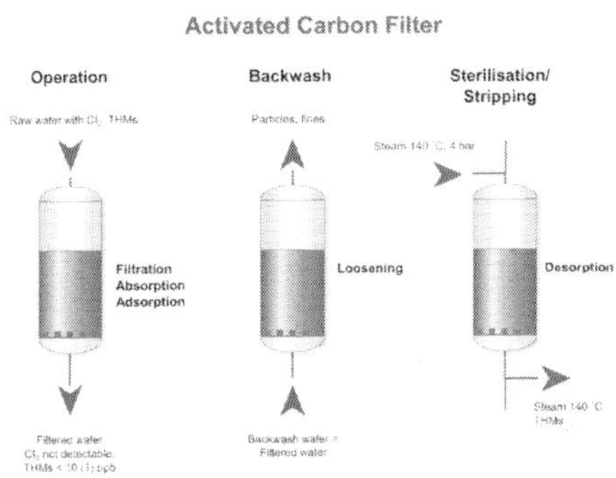

Figure 3.2: The use of carbon during the absorption process is commonly used in the filtration systems.

Source: https://www.euwa.com/activated-carbon-filtration.html.

In matters carbon filter systems, there are two major carbon filter systems. There is one that makes use of granulated carbon while the other makes use of solid carbon.

As mentioned above, the filtration process makes use of activated carbon. The activated carbon is achieved through a process known as heat treating. The process results to an activated carbon having microscopic pores. They also have fissures that will trap both chemical and organic impurities. The process in a way mimics the filtering properties of the earth.

For this technique to be effective, there are some factors that come into play. The carbon filter system is dependent on the type of carbon used, amount of carbon used, the design of the filter system and the duration in which the filter system has been used. It is also dependent on the particle size of the carbon filter and also the rate at which the liquid is flowing through the filter. This is applicable in the sense that if the fluid flows through the filter at a slower rate than the exposure time of the contaminants to the filter will be longer resulting to water which contains less amounts of impurities. The filter system is known to remove contaminants such as hydrogen sulfide, chlorine, Radon, volatile organisms, benzene, odors, colors, and pesticides. For instance, the charcoal carbon filters are effective in removing chlorine elements, sediments, taste, odor, and also volatile organic compounds (Singh, 2006).

Mechanism Used in Carbon Filtration: The mechanism focuses on absorption. It involves the contaminants in the fluid being treated but first, they are trapped within a pore structure made of carbon substrate. The carbon substrate itself is made of carbon granules that are said to be relatively porous. This leads to the substrate having a relatively large surface area where the contaminants are caught. In the case where activated carbon is used, it provides a larger surface area compared to using non-treated carbon. This attribute also allows it to be used in a number of applications.

Based on the type of carbon used, there are majorly two types of carbon filtration. They include:

Bed Filtration: This type of carbon filtration employs the use of granulated carbon. By using this method there are two choices availed to the water treatment plant. They get to decide either to use an existing filter or change and use a new filter. Before making the decision, they need to look into factors such as water flow and bed depth among others. When using this method, the operators will be required to make (on a given time frame) the changes in water therefore the backwash of water may be conducted. The benefit of using granulated carbon it that it can be reactivated. This is made possible through the use of thermal oxidation. It can therefore be used a number of times. The granulated carbon can also be used on its own or conjoined with other technologies so that the desired water quality is attained. The most common is it being used with ozone treatment resulting to the biologically activated carbon process (Talwar, 2012) (Figure 3.3).

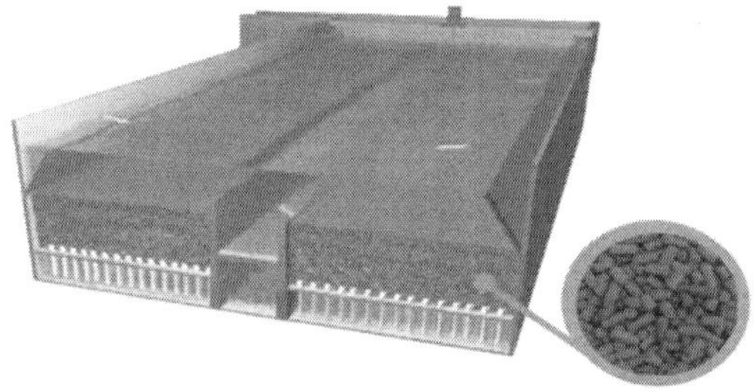

Figure 3.3: Showing granulated carbon filtration.

Source: https://www.suezwaterhandbook.com/degremont-R-technologies/drinking-water-production/treatment-on-activated-carbon/granular-activated-carbon-contactor-and-filter-Carbazur-R.

Secondary Filtration: This method also involves the use of granulated carbon. Involves them being placed in treatment plants. They are incorporated after the filtration process or after flocculation and sedimentation. These granulated activated carbons are known to remove contaminants such as chlorides, heavy metals, unwanted water disinfectants, tannins, trihalomethanes, and toxins generated by algae. This method of purification is mostly used in villages whose water has been contaminated by perfourionated compounds. For instance, in Florida, this method is used in removing contaminants such as dieldrin.

Advantages of Using Carbon Filtration: Carbon filtration is beneficial when used as it allows the removal of dissolved organic compounds. It is also known to effectively remove some chemicals. The other benefit is that this method of filtration can be combined with other water purification techniques. This is attribute to the fact that it works well together with other filtration systems. The other advantage is that it removes both objectionable flavors and odors. The other advantage is that it can be used in chlorine absorption therefore its effective removal from water (Johnson, 2006).

Disadvantages of Using Carbon Filtration: Some of the disadvantages of this method include the fact that it may not efficiently remove heavy metals present in water. This means that the resultant water may not be entirely pure. Also, this method requires that the filters be changed at a given frequency. This is important as it contributes to the filtering capacity. This method is also not effective in the removal of salts, minerals, and dissolved inorganic substances (Johnson, 2006).

Apart from Carbon Filtration, there are other methods of filtration used in water purification. Among them is the rapid sand filters.

3.4.2. Rapid Sand Filters

While using this filter, water is made to move vertically through the sand. Before it reaches the sand, there is usually a layer of activated carbon or anthracite coal present above the sand. The various layers play a variety of roles. For instance, the first layer is used to remove organic compounds known to contribute to odor and taste. There is a space present between the sand particles which is relatively large when compared to the suspended particles. It creates the need for one to conduct a more comprehensive filtration process. Though some of the particles may pass through the surface layers, they end up being trapped in the pore spaces or the sand particles. For an efficient filtration to be conducted, the filtration may have to extend into the depth of the filter. The depth of the filter is a significant property as it plays a crucial role in its operation. This is relevant in the case where if the top layer of sand was to effectively block the particles then the filter will not be operational (Figure 3.4).

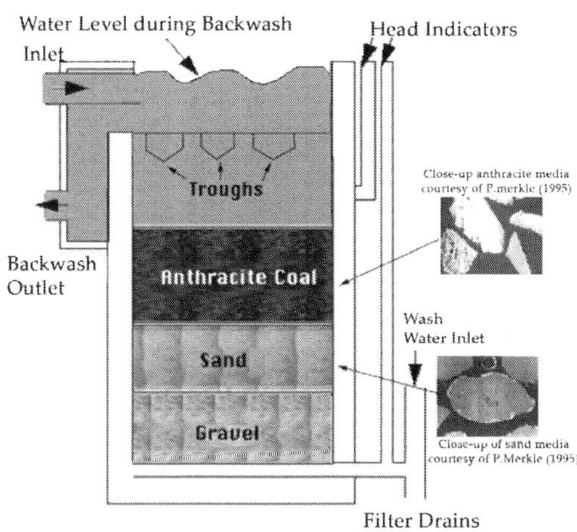

Figure 3.4: Showing rapid sand filters.

Source: http://www.elaguapotable.com/WT%20-%20Rapid%20Sand%20Filtration.htm.

This will require that the filter should be cleaned. Cleaning the filter involves water being passed quickly in an upward manner through the filter and hence the name backflushing. This is done to remove the embedded particles. Also, compressed air may be blown up at the bottom of the filter. This is done to break up the compacted filter, therefore, making the backflushing process to be much easier. This process is referred to as air scouring. The resultant water from the backflushing process is disposed of. Accompanying it may be the sludge produced in the sediment basin. There is also the option of recycling it. This is achieved by mixing the raw water that is entering the plant. This practice is discouraged as it may lead to the re-introduction of concentrated bacteria to the raw water (Talwar, 2012).

For some water purification plants, they opt to use pressure filters. These filters employ the same principle used by gravity filters. The difference may be that the filter medium in this case is enclosed in a steel vessel. The water is therefore forced through it under pressure.

Advantages of the Rapid Sand Filters: These filters have the ability to filter out relatively smaller particles when used in comparison to paper and sand filters. They are also able to filter out particles that are larger than their pore sizes. The rapid sand filters are relatively thin. This allows liquids to flow through them at a rapid pace. These filters are also strong and therefore have the ability to withstand pressure across them that are within the range of 2–5 atmospheres. The other advantage is that they can be reused. This is made possible through the backflushing process.

3.4.3. Slow Sand Filters

Slow sand filters are among the water purification techniques. It is, however, practical in areas where there are land and space allowing the water to slowly flow through the filters. The filters used in this technique are dependent on biological treatment processes compared to the physical ones. The construction of these filters involves layers of sand that are graded. It is made in a manner that fine sand is at the top, there are layers of both coarse sand and at the bottom is gravel. The drains present at the base collect water that is to undergo a disinfection process. Filtration is dependent on a number of factors. It includes the development of thin biological layers also referred to as the zoogleal layer which may be present at the surface of the filter (Figure 3.5).

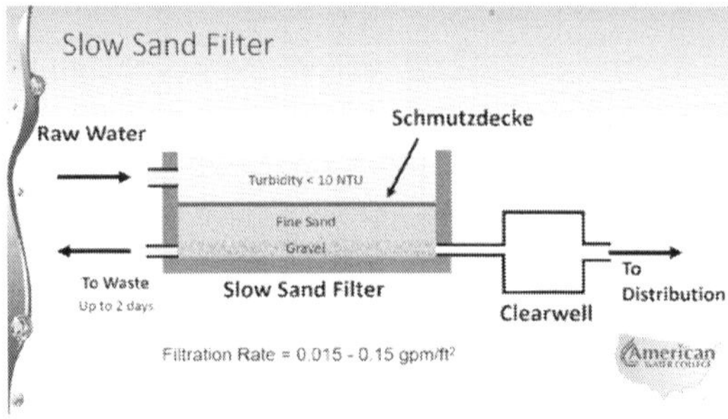

Figure 3.5: Slow sand filters.

Source: https://www.youtube.com/watch?v=fBKlXDx6txo.

For effective slow sand filters, they can be used over long periods of time. This is dependent on the design of the pretreatment. If it is well designed, then the produced water may have a low composition of nutrients. The low level of nutrients facilitates the safe movement of water while in the distribution system. They will have low disinfectant levels resulting in a lower chance of consumer irritation due to chlorine and its products. Different from the rapid sand filters, the slow sand filters are not backwashed. Their maintenance involves the scrapping off of the top layer of sand. In the case of large scale use, the slow sand filter will be referred to as bank filtration. This is because the naturally occurring sediments on river banks are used in the provision of a first stage contamination filtration. The water produced as a result of this process may not be safe for drinking.

3.4.4. Membrane Filtration

This method is commonly used in water purification processes involving both sewage and drinking water. In the case of drinking water, the membrane has the ability to remove particles larger than 0.2 μm. These filters are efficient when employed in a tertiary treatment. This is in the case where one wants to reuse water in industries, before releasing water into rivers that are used by the population or if the water is meant for domestic purposes. This method is mostly used in industries more so those involved in beverage preparation (Figure 3.6).

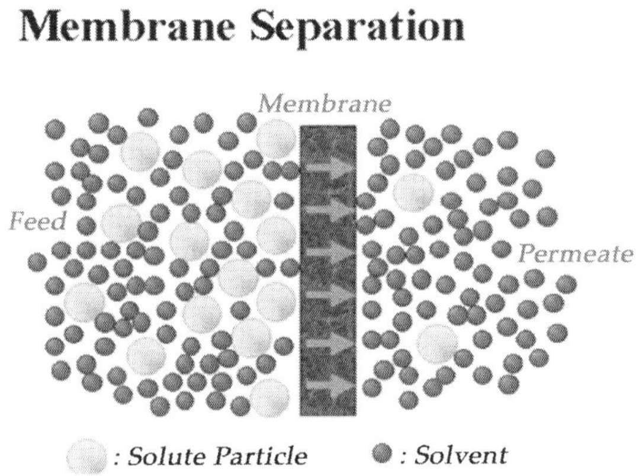

Figure 3.6: Showing membrane filtration.

Source: https://canvas.jmu.edu/courses/1484140/pages/membrane-filtration.

The major disadvantage of using the filtration process is that it may not have the ability to remove contaminants that have dissolved in water. Under filtration, there is a method referred to as ultrafiltration which is similar to filtration process that employ the use of microfilters.

3.4.5. Ultrafiltration

This system of water purification involves the removal of contaminants by allowing water to flow through a screen. The screen is used to refer to a semi-permeable membrane that is known to efficiently separate molecules present in water. It, therefore, allows certain molecules to pass through it while others are retained. This method removes sediments, microorganisms, and extremely fine particles. The advantage of using this method is that a significantly good quality of water is produced while a low energy investment is employed. The disadvantage may be that there are chances of the membrane being clogged. This method can also be relatively expensive in the case where there is the presence of large amounts of sediments. This method does not remove dissolved particles (Cheremisinoff, 2002).

3.4.6. Candle Filter Purification

This method of filtration employs the use of a candle filter. This type of filter has basic operating mechanisms. It involves the use of candles which have a relatively small or minute pores. Therefore, any particle that has a size larger than the candle pores is blocked and clean water gets to penetrate through. This filter does not require any type of electricity for it to run. The disadvantage of this filter is that it is unable to prevent the movement of microbes. This requires that the water should be boiled before one consumes it. It is also required that the candle should be cleaned up on a regular basis to allow for effective operation.

3.4.7. Distillation

This method involves water being heated to boiling. The water is then cooled so that the water vapor is condensed. The impurities present in water are filtered out through evaporation. The condensed water is then stored and afterward collected in a glass-lined container. The water produced from this process is rarely used for drinking. This is attributed to the fact that the water produced has low amounts of oxygen and also the taste is termed as flat. This method of water purification is known to remove bacteria, sodium chloride, pesticides, organic chemicals, nitrates, and also lead. The advantage of using this method is that it removes a relatively wide range of contaminants. It is also re-usable. The disadvantages include that this water purification process is relatively slow. It also requires a large amount of energy and water. This method also leads to the concentration of the contaminants due to the high boiling points (Figure 3.7).

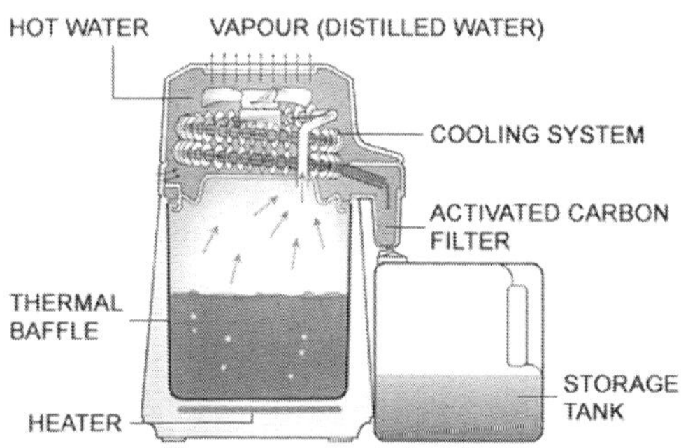

Figure 3.7: Showing distillation.

Source: https://www.pure-pro.com/water_distillers.htm.

3.4.8. Ion Exchange

This method involves the removal of metallic elements present in water. The method can at times be referred to as Water Softening. This method makes use of a large tank that is filled with a special negatively charge resin. The resin beds used in ion exchange serve as the basis for the ion exchange process to be conducted. Water is made to pass through the resin. As water passes through the resin bed, the metallic ions (carrying a strong positive charge) displace the weaker charged Sodium and potassium ions. Therefore, the metallic ions are trapped as a result of the electromagnetic attraction to the resin particles (Figure 3.8).

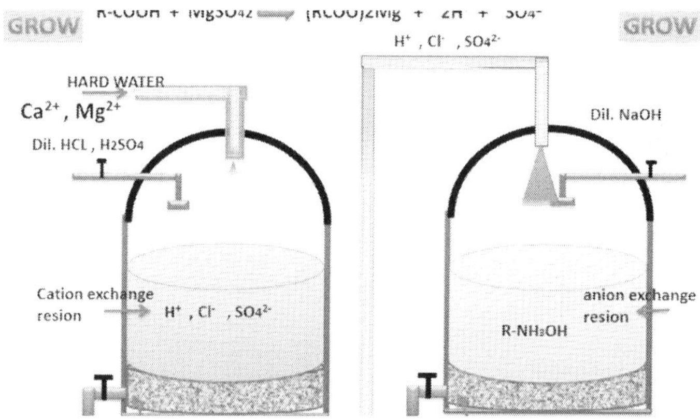

Figure. 3.8: Ion exchange process.

Source: https://www.youtube.com/watch?v=BmpknJNDXfE.

The resin beds used in ion exchange has a system that allows an automatic cleansing. The beds are also regenerated at given intervals depending on the water volume. The regeneration process in this case involves the beds being flooded with a supersaturated saline solution. By so doing, there is an effective sweeping away of the metallic ions from the resin bases. Ion exchange as a method of water purification facilitates a highly effective removal of metals known to be responsible for pipe scaling and deposit. This process also involves the removal of other heavy metals including mercury, iron, lead, and cadmium. These metals have been said to be associated with the commonly known health concerns (Johnson, 2006).

The resin beads have a coating of sodium exchange. The method is also known to be highly effective when used in conjunction with other techniques such as reverse osmosis and carbon filtration. It is important to note that when using this method, it is crucial that the metals are removed in the initial purification process. This ensures that the sensitivity of the technology is used in other stages. This method is used in removing nitrates, fluoride, iron, sulfate, manganese, calcium, and magnesium.

The advantages of using this method is that it removes dissolved inorganic minerals and metals effectively. This method is also less expensive. The disadvantages include the fact that it does not effectively remove bacteria. Also, the beads have the ability of trapping bacteria and they can be possible breeding grounds (Johnson, 2006).

3.4.9. Ultraviolet (UV) Radiation

This method involves the use of low level UV light. This light kills both bacterial as well as viral bacteria. As dealing with radiation may be dangerous, this method involves the production of a negligible amount of radiation therefore no traces are left behind. It may involve the use of a typical UV sterilizer that looks like a stainless cylinder. There is a water inlet on one end of a cylinder and also water outlet on the other end of the cylinder. The method removes both viruses and bacteria (Figure 3.9).

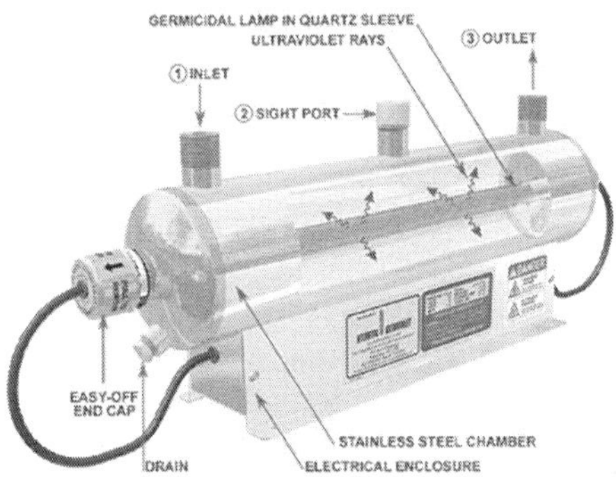

Figure 3.9: Showing ultraviolet radiation water purification.

Source: https://watersafe.com.pk/index.php/tag/water-purification/.

This method of purification is different from others as it does not employ the use of chemicals and is able to produce water that is completely free from microbiological contamination. This method is also used in facilitating the maintenance of the sanitized water. UV light was considered for water treatment as it is highly effective in activating cysts present in low turbidity water. Therefore, the UV light's disinfection effectiveness reduces as the water turbidity increases. This is attributed to absorption, scattering, and shadowing caused by suspended solids. Apart from UV light. Ionizing radiation such as X-rays, gamma rays and electron beams can be used in the sterilizing of water (Talwar, 2012).

The method involves water being passed through a special chamber that contains an UV source. The UV light in this case is the sterilizing agent. At a particular wavelength, the bacteria, viruses, and other microbial contaminants in the water would lead to the destruction of the genetic material in the organisms. This leads to the elimination of the chance of both a bacterial or viral reproduction and proliferation occurring. The organisms are killed and are captured. They are also removed by conducting a pre-filtration. This is done in the case where a reverse osmosis purification is to be conducted.

There are control technologies present throughout the process. This is availed as there may be times that the efficiency of the UV lamp may deteriorate over time. There is a device present in the UV lamp that measures the intensity of the radiation produced by the UV lamp. UV water purification is commonly used in conjunction with other techniques of filtration such as reverse osmosis or activated carbon filtration.

3.4.10. Disinfection

This is also among the techniques used in water purification. It is achieved through conducting a filtering as well as the addition of disinfectants. The disinfectants used in this case are mostly chemicals. The water is disinfected so that the pathogens that are present in it are killed. The disinfection occurs when the pathogens pas through filters. There is a residue dose of disinfectant that kill or inactivate potentially harmful microorganisms that may be present in the storage and distribution systems. Disinfection may be used in removing both viruses and bacteria such as cholera and also protozoa.

Once the chemical disinfectant is introduced to the water, the water is temporarily held in storage which is also known as a contact tank to facilitate the disinfection action to be done (Figure 3.10).

Chlorine Disinfection: It is also referred to as chlorination.

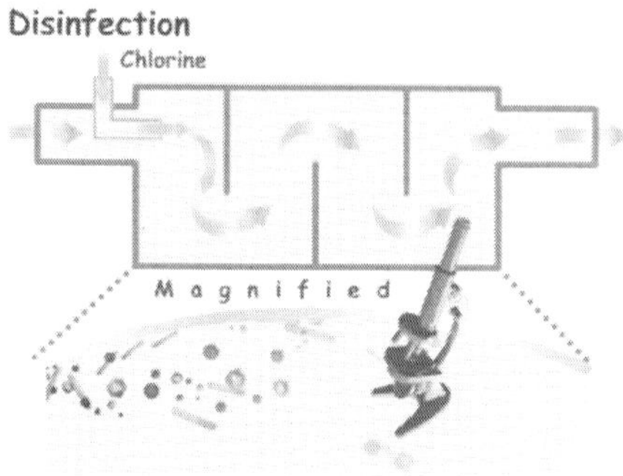

Figure 3.10: This method is the most common as it is used in most households. It involves the use of chlorine compounds such as chloramine or chlorine dioxide.

Source: https://jzcacayuran.wordpress.com/2015/07/22/water-treatment-process/.

Chlorine is commonly used as it is a strong disinfectant as it has the ability to kill many harmful microorganisms. This is attributed to the fact that it is a toxic gas. It therefore requires one to be careful when using it. Though when chlorine gas is used there is a chance that it may be released to the atmosphere, therefore manufactures have looked into using sodium hypochlorite which is relatively inexpensive and releases free chlorine when used. Chlorine solutions can also be generated through conducting electrolysis of common salts. A solid form of calcium hypochlorite also releases chlorine when added in water. Human handling of this solid requires a more frequent contact the process is therefore automated (Johnson, 2006).

The generation of sodium hypochlorite in liquid form is relatively inexpensive and safer compared to using chlorine gas or solid. There are given levels in which chlorine when present in water is said to be safe for consumption. The disadvantage of using chlorine is that it has the ability to react with natural organic compounds present in water. Once combined, there is a possibility of them forming harmful chemical by-products as some of them are carcinogenic in large quantities. The recent approach involves the initial removal of the organic chemicals before chlorine or its products are added. Though chlorine has been said to remove bacteria present in water, it is not sufficiently effective in removing pathogenic protozoa that may lead to the formation of cysts in water (Cheremisinoff, 2002).

Chlorine Dioxide Disinfection: Using chlorine dioxide in water purification is much better than using chlorine on its own. However, it is rarely used as it has been attributed to the creation of large amounts of chlorite. Chlorite is a by-product that should be regulated to attain low level amounts. The chlorine dioxide may be supplied as an aqueous solution. This is advantageous as it may be added to water and thereby minimizing the risk of handling gas. If chlorine dioxide gas accumulates then there is a possibility of it spontaneously denoting.

Chlorimination: This type of water disinfection involves the use of chloramine. This is becoming more common amongst the population. Though chloramine is not a strong oxidant, it is significantly useful as it provides a longer-lasting residual compared to free chlorine. This is attributed to the fact that it has a lower redox potential when compared to free chlorine. The benefit of using it is that it does not combine with organic compounds leading to the production of by products that may be harmful to human health. Chlorine can be converted into chloramine through the addition of ammonia into the water. This is done after chlorine has been added. For the water distribution system that are disinfected using chloramine, they may encounter nitrification. This is because ammonia is a nutrient for bacterial growth. The nitrates may be generated as a by-product.

Ozone Disinfection: It is also referred to as ozonation. The ozone is an unstable molecule and to attain stability, it readily gives up one of the oxygen atoms therefore providing a powerful oxidizing agent that tends to be toxic to the waterborne organisms. This type of disinfectant is strong and has broad uses. It is mostly utilized in European nations. In matters inactivating harmful protozoa, this method has proven to be effective. It has led to the minimization of the production of cysts. It is also used in the killing of pathogens. Ozone is created by passing oxygen through UV light. For it to be utilized as a disinfectant, it is required that the ozone be created at the site and afterwards be added to water through double contact.

When used, it results to production of fewer dangerous by-products. It also eradicates problems such as bad taste and odor. It is also beneficial in that there is no ozone that is left in the water after the disinfection is conducted. In the case where there is no residual disinfectant in the water, then chlorine or chloramine may be added in the distribution system to allow the removal of potential pathogens in the piping.

Ozone has been successfully used in water plants in most countries. It is termed to be safe. In most of the plants, it is used as an anti-microbiological agent during the treatment process. It is also used in water storage as well as food processing. The recent research shows that ozone reacts with bromide ions leading to the production of concentrated carcinogen bromates. As bromide ions may be found present in freshwater supplies in relative concentrations it may lead to some plants opting not to use it. Ozone disinfection is energy-intensive (Singh, 2006).

Bromination and Iodization: Both bromine and iodine can be utilized as disinfectants. They are rarely used as chlorine is said to be three times more effective as a disinfectant. Iodine is mostly used in portable water purification while bromine is used as a swimming pool disinfectant.

Portable Water Purification: This is among the recently developing trends under methods of water treatment. Involves the use of devices and methods present for the treatment and disinfection of water during emergencies. In this case, disinfection is the major priority. Other factors such as taste, appearance, and odor and trace chemical contamination are said to not affect the short-term safety of the drinking water.

3.4.11. Reverse Osmosis

This technique applies the same principle used by our body's cell which involves water being filtered through a semi-permeable membrane present between a concentrated and diluted solution. This involves two solutions having different concentrations are being separated through a semi-permeable membrane. The solution having a higher concentration will enter the lower concentrations. This process may be reversed by the application of pressure on the sides that holds the more concentrated solution. This forces the water through the membrane thereby removing impurities from water.

For a successful reverse process, the membrane used should be of high quality. The quality of the produced water is therefore dependent on the quality of membrane used. The membranes used in reverse osmosis systems should be able to trap salts, inorganic solutes and other organic compounds. Reverse osmosis is known to provide high-quality drinking and cooking water. The volume of water required to produce a usable amount of water may make it quite impractical for it to be used in whole-housing filtering. The reverse osmosis is commonly combined with other techniques used in water filtration to facilitate a complete water filtration solution (Singh, 2006) (Figure 3.11).

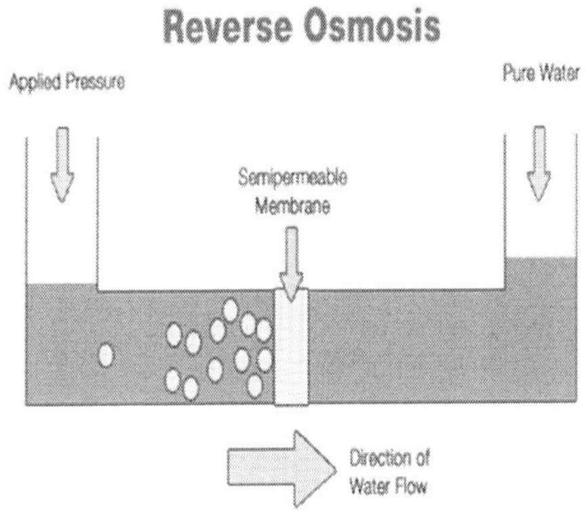

Figure 3.11: Reverse osmosis is used in removing contaminants such as lead, calcium, magnesium, sulfate, sodium, manganese, potassium, chloride, aluminum, fluoride, nitrate, boron, microorganisms, and other organic contaminants.

Source: https://www.worldofchemicals.com/610/chemistry-articles/reverse-osmosis-water-treatment-process.html.

3.4.18. Desalination

This method is mostly used when the water is saline.

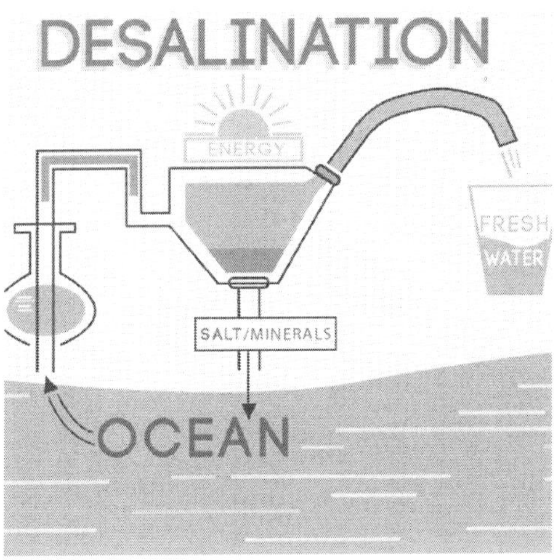

Figure 3.12: The saline water is converted to freshwater.

Source: https://tradeexworldsourcing.com.au/efficient-ways-to-desalinate-water/.

This method is used in treating seawater. This method incorporates other methods of water purification such as distillation and reverses osmosis. Desalination of water purification is actually expensive when placed in comparison to other methods of water purification. Also, it is not commonly used by populations as they are not satisfied with water produced from distillation. Desalination is economically practical for high-valued uses. For example, it is suitable for industrial applications which require purified water, such as the production of electronic parts, special foods and pharmaceuticals, and uses reverse osmosis as a component of the production process in which the concentration and/or fractionation of the wet process stream is necessary . Farmers, including agricultural uses for greenhouses and hydroponic methods, may also use the process to desalinate and purify water for greenhouse irrigation (besides, desalination water is low in microbes and nematodes, which also aids in controlling plant diseases). (Figure 3.12).

3.4.13. Centrifuge Method

This method involves the use of gas hydrate crystals. In a case where carbon dioxide or any other low molecular weight gas is mixed with water having contaminants, gas hydrate crystals will be formed exothermically. Once this is done, the separation of the hydrated crystalline is achieved through sedimentation or centrifuge. If sedimentation is done, it is usually accompanied with decanting. Water is then released from the hydrate crystals through heating.

3.4.14. Bio Remediation

This method involves microorganisms being used in removing certain waste products in a given area. This method has been in use since 1991 and is used in removing impurities from water such as metals, alkanes, and perchlorates. This method in most cases is used in the treatment of ground water. The method is done with relation to perchlorate as well as chloride compounds (Figure 3.13).

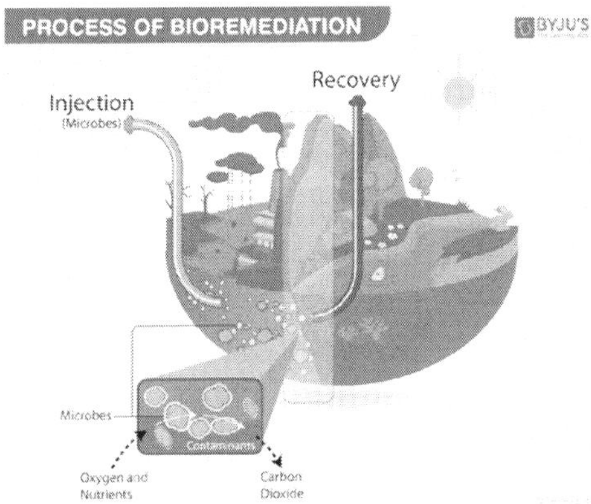

Figure 3.13: Bioremediation has been proven to be more effective to perchlorate compounds are relatively soluble in water.

Source: https://byjus.com/biology/bioremediation/.

This implies that it is difficult to remove it from water. Although this method is significant, the implementation has not been feasible as there is still more studies that need to be conducted on regards to the rates. They also look into the effects of the microbial activity and also the production of large scale implementation of the method.

3.4.15. Direct Contact Membrane Distillation

This method is mostly effective when done in conjunction with distillation. It involves heated seawater is made to pass along the surface of a hydrophobic polymer membrane. The evaporated water is then made to pass, from the hot side, through the pore in the membrane into a stream of cold pure water that is present on the other side. The difference in vapor pressure present between the hot and cold side assists in pushing water molecules through the membrane.

3.4.16. Granular Activated Carbon Adsorption

This method involves the use of granulated carbon that is activated. The granulated carbon in this cases tends to have a large surface area. It plays the role of adsorbing most of the compounds such as toxic compounds. Water is made to pass through activated granulated carbon. This method is employed in municipal regions having organic contamination, odors, and taste. Most of the household water filter systems make use of the carbon filters in an aim of purifying the water. For some of the filters, they can contain impurities such as silver presenting itself as a silver nanoparticle. In the case where silver is held in the carbon block for a long period of time, it may result to microorganisms may grow inside. This will lead to fouling and contamination. The silver nanoparticles are efficient in anti-bacterial material. They have the ability to decompose toxic halo-organisms' compounds. A good example may include the conversion of pesticides into non-toxic organic products.

When using this method, the filtered water must be used immediately it is filtered. This is done because the low amounts of the residual microbes may end up proliferate within a given period of time. This method is used in most homes and it is able to remove 90% of chlorine present in a glass treated water. It is essential that the filters be periodically replaced. This is important because if it is not done the bacterial content in the water may increase as a result of the growth in bacteria.

3.4.17. Boiling

This method is used in many of households. It is actually championed to be a common practice to be done in most of the households, especially when municipal, or other sourced water is not safe to drink or is suspected to contain microbiological contaminants harmful to human health. This method is cheap and to a greater extent effective for use. This method involves boiling water to its boiling point which may vary depending on the altitude of the area. In most cases, the boiling point is 100 degrees Celsius. This method is considered to be a traditional mode of water purification. It is efficient in treating most of the microbes present in water that have been known to be the cause of most intestinal related diseases. This method to some extreme is disadvantageous because it is not able to remove the chemical toxins amongst other impurities present in water (Singh, 2006).

There also some microbes that a resistant to heat but most of them do not result to intestinal diseases and are therefore harmless. This method involves the boiling of water for about ten to ensure the thorough killing of the microbes this is because most of the microbes are unable to survive in temperatures above 60 degrees. In cases where the water is hard water, when boiled there is a decomposition process that occurs. In this case, the bicarbonate ions decompose causing a partial precipitation such as calcium carbonate. This method does not remove solutes that have a higher boiling point that is higher than water. Boiling could also lead to the increase in concentration of the contaminants. The use of this method is advantageous as there are no residual disinfectants in water. Water when boiled, is not really advisable for storage of long periods of time as there are high chances of the water acquiring new pathogens (Singh, 2006).

3.4.18. Municipal Water Treatment

This is a common method used in large scale purification of water. The water from treatment plants is the one supplies to most of the households in the cities. The main aim of water treatment is to remove unwanted contaminants present in water this will make water safe for drinking and can also be used in industrial processes as well as medical applications. This method involves a variety so that the water will be clean and safe for use. This method is able to remove contaminants minutes such as fine solids, microorganisms, and also dissolved organic and inorganic materials. The methods employed in the treatment plants are dependent on the quality of water, the cost of the treatment and also the expected standards of the quality of water expected after the water is processed. There are a variety of processes involved in the treatment plant.

Pretreatment: This involves an outline of activities that are done before the water undergoes the treatment process. The first step involves pumping containment. This involves most of the water being pumped from the point source and it is then transported into the holding pumps or pipes. It is important to ensure that the pipes and infrastructure should be well maintained to reduce the chance of the water being further contaminated. This therefore may lead to the minimization cases of accidental contamination. The next step is screening. This involves the removal of contaminants such as large debris. The debris includes sticks and leaves among others. They are removed because if they are present in future processes it may interfere with the operation. However, this method is not used for water such as groundwater (Cheremisinoff, 2002).

The next step is storage. The water from the sources can be stored in bank side reservoirs for a given period of time. This is done to allow certain biological purification to occur. This step is applicable in the case where the slow sand filters are used. The storage reservoirs facilitate the buffer against a short period of drought. This allows the water supplies to be maintained in case of pollution incidents. The final step is pre-chlorination. This involves the incoming water being chlorinated to minimize the growth in the fouling organisms in the pipe-works as well as tanks.

Water treatment involves the following:

pH adjustment: In the case of pure water, it tends to have a pH of about 7. They are neither alkaline nor acidic. For instance, the seawater usually has a pH that ranges from 7.5 to 8.4 and is generally referred to as moderately alkaline. On the other hand, freshwater has a variety of pH values and is dependent on the geology of the aquifer or drainage basin and the influence of the contaminant inputs. In most cases, if the water is acidic then it will have pH to improve the pH one can add lime, soda ash, or sodium hydroxide. For instance, if lime is used in water having calcium ion concentration, it may lead to an increase the water hardness. In the case where the water is highly acidic water, detoxification has been proven to effectively raise the pH. This is done by stripping dissolved carbon dioxide from water.

The making of the water alkaline helps coagulation and flocculation processes so that they can work effectively. It also assists in the reduction of the risks of lead being dissolved from the lead solder and lead pipes in pipe fittings. Having a significant alkalinity will minimize the corrosiveness of water to iron pipes. Acid could be added to alkaline waters in some situations to minimize the pH. The ability of water precipitating calcium carbonates to protect metal surfaces and minimizing the ability of toxic metals being dissolved in water which is a function of pH, temperature, alkalinity, mineral content and calcium concentration.

Coagulation and Flocculation: The initial step in a conventional purification process involves the addition of chemical that will aid in the removal of particles that have been suspended in water. The particles in the water can be inorganic. They include clay and silt. They can be organic which includes algae, viruses, bacteria, protozoa, and natural organic matter. Both the organic and inorganic particles may be a major contributor to the color and turbidity of water. Adding inorganic coagulants including aluminum sulfate or iron salts may result to simultaneous physical and chemical interactions on and among the particles. After a short period of time, the negative charges on the particles are neutralized by inorganic coagulants. Also, the metal hydroxide precipitates of iron and aluminum ions start to form. The resultant precipitates merge into larger particles. This is done under natural processes like Brownian motion. It can also be done through induced mixing and therefore termed as flocculation (Talwar, 2012).

The amorphous metal hydroxide is called Floc. The large amorphous aluminum and iron (ll) hydroxide will adsorb and the enmesh particles present in the suspension. It facilitates the removal of particles through subsequent processes such as sedimentation and filtration. The aluminum hydroxide that are formed are usually within a relatively narrow pH range. Iron (ll) hydroxides can be formed over a relatively large pH range that includes pH levels that are lower. They are also effective for alum. There are various conversations that have been sparked up in terms of the usage of coagulation and flocculation (Talwar, 2012).

In the water plants, there is utilization of high energy, the quick mix unit process which involves coagulant chemicals are added. It is then followed by flocculation basins. This involves low energy inputs turning large paddles. Coagulation and flocculation processes are conducted once the metal salt coagulant is introduced. The organic polymers were created in 1960s to assist coagulants. In other cases, it may be used as replacements for the inorganic metal salt coagulants. In the case of synthetic organic polymers have high molecular weight compounds that contain particles. These high molecular weight compounds adsorb onto the particle surfaces as well as through inter-particle bridging coalesce with other particles to form floc. A common cationic organic polymer is POLYDAMAC. This polymer is used in water purification plants.

Sedimentation: The water leaving the flocculation basin may be allowed to pass through the sedimentation basin. The sedimentation basin is also called a clarifier or a settling basin. It involves the use of a large tank having low water velocities therefore allowing the floc to settle to the bottom. It is crucial that the sedimentation basin is best located nearer to the flocculation basin. This is done so that the transit between the two processes may not allow settlement or floc break up. The sedimentation basins can be rectangular. The water flows from end to end. The sediment basin can also be circular in the case where the flow is from the center outward. The outflow in the sedimentation basin is basically over a weir so only a thin top layer of water. The furthest from the sludge will exit (Johnson, 2006).

In 1904, the efficiency of the sedimentation process was demonstrated as a function of the particle settling velocity. It is also a function of the flow through the tank as well as the surface area of the tank. The sedimentation tanks are designed within a range of overflow rate of given gallons per minute per square foot. Generally, the sedimentation basin efficiency is actually a function of the detention time or the depth of the basin. However, it is important that the basin depth be sufficient so that the water currents do not disturb the sludge as well as the settled particle interactions are promoted. As the particle concentrations in the settle water increase near the sludge surface on the bottom of the tank. The settling velocities may increase as a result of the collisions and agglomeration of the particles (Johnson, 2006).

The typical detention durations for sedimentation may vary from 1.5 to 4 hours while the basin depths vary from 10 to 15 feet. In the case of traditional sedimentation basins, inclined flat plates or tubes may be added. This is done to improve particle removal performance. The inclined plates and tubes may lead to the drastic increase in the surface area available for the practices to be removed in concert with the Hazen's original theory. For the given case scenarios, the amount of ground surface area occupied by a sedimentation basin having inclined plates or tubes can be relatively smaller than a conventional sedimentation basin.

Sludge Storage and Removal: As the particles from the water settle at the bottom of the sedimentation basin, it leads to a layer of sludge being formed at the floor of the tank. This floor must later on be removed and treated. The amount of sludge that is generated should be significant. This implies that 3 to 5% of the total volume of a water treatment plant. The sedimentation basin can be equipped with mechanical cleaning devices that may lead to a continuous cleaning of its bottom. It may also involve the basin being periodically taken out of service as well as the manual cleaning.

Floc Blanket Clarifiers: This is a subcategory of the sedimentation process. It involves the removal of particulates through entrapment in a layer of suspended floc as the water is being forced to move upwards. The greatest advantage of the floc blanket clarifiers is that they occupy a relatively smaller footprint compared to the conventional sedimentation. The downside is that the particle removal efficiency may be highly variable. It is therefore dependent on the change in the influent water quality and influent water flow rate.

Dissolved Air Flotation: As the particles that are to be removed do not settle out of solution easily, it has led to the need for a dissolved air flotation. Therefore, after the coagulation and flocculation process, the water is made to flow through the DAF tank. Here, air diffusers at the bottom of the tank creating fine bubbles that attach to the floc resulting to the floating mass of concentrated floc. The floating blanket is then removed from the surface. The clarified water is then withdrawn from the bottom of the DAF tank. In this case, the water supplies are usually vulnerable to the unicellular algae blooms and supplies having low turbidity and a high color of the water may involve the use of DAF.

After this is done, there are other processes that can be conducted which include filtration among others. In matters related to water purification technology, there is the issue of safety and controversy.

3.5. SAFETY AND CONTROVERSY

This involves the evaluation of the water purification techniques depending on the basis of how safe is the resultant water after they are subjected to the technologies. For instance, chlorine has been said to pose a danger in terms of its handling and it may be accidentally released to the environment. Therefore, it has led to people opting to use chloramine as the disinfectant agent. The safety also looks into matters of the minerals present in water. Some of the minerals present in water are actually useful for the body. There are technologies such as distillation which results in the total removal of minerals from water. This, therefore, creates the need for these processes to be done in conjunction with other processes so that some of the minerals are returned to the water.

CHAPTER 4

Air Purification Technologies and the Environment

CONTENTS

4.1. INTRODUCTION

Air purification systems as much as it may appear to be a new concept has a very fascinating and long history. Many people may assume that because the advanced technology is relatively recent that it is a purely contemporary idea, however, it dates back to the time of the Industrial Revolution. The Industrial Revolution is very important in the formation of air purifier technology in the sense that it heralded more economic opportunities for citizens and mass-scale production however it came at a very heavy cost of polluted air. The Industrial Revolution was characterized by the use of coal as the primary power source to provide heat and run all sorts of machinery but it is not a fuel that burns clean at all.

Coal actually starts out as peat which is decomposed organic matter that accumulates on the earth surface and over millions of years is buried under heat and pressure and forms lignite, sub-bituminous, bituminous, and anthracite coal. Coal is ordinarily found within layers of rock that have compacted and folded into mountains and is an abundant resource throughout the planet although they are not evenly distributed. The largest deposits of coal are found in the US, China, Russia, India, and Australia. It is a mineral that requires little to no refinement before it is burned as fuel.

Coal contains sulfur, nitrous oxides, and other dangerous metals like mercury lead and arsenic that are a byproduct released into the air when it is burned. Cool also releases a large amount of particulates and carbon dioxide that are a danger to one's health and increases air pollution. The number of airborne toxins and pollutants are numerous just like with very many other fossil fuels which are limited and non-renewable. Over time the slow unproductive economies turned into huge successes but noticed ubiquitous clouds of smoke and many began to suffer because of the concentrated amounts of polluted air across many cities in the United States, Britain, and continental Europe (Xu, 2018).

The story begins with John and Charles Deane two brothers from Deptford, South East London, England. They studied at the Greenwich Hospital School for Boys which was primarily a school for naval pursuits, with Charles having left as a Seaman at the age of 14 for seven years before returning to Deptford. In his time employed as a call car he recognized the many problems of fighting fires within the walls of ships. John Deane was in England and in the 1820s was trapped in a stable that had gone up in flames. He realized the only way that through the flames and the fumes was to put on a medieval Knight in armor helmet and pumped in air through a hose from a fire brigade water pump. He managed to survive an even better yet he rescued all the horses. In 1823, Charles Deane patented the smoke helmet and created it as the first mask four firefighters allowing them to enter buildings that were burning without being overcome by the fumes because it filtered toxic air. The technology was improved upon in 1854 by John Stenhouse who introduced a charcoal-based filter design that works very close to how carbon filters in the air purifier industry do today by blocking even more noxious gasses. Not much later John Tyndall made his upgrade by introducing a respirator to assist breathing and a combination of the respirator device together with the more advanced filtration systems became early forms of today's respirator masks used by divers and miners (Xu, 2018).

These key advancements were made because of the realization that occupants' comfort and health can be adversely affected by poor air quality. Research carried out in 2016 by the BBC reported that more than 5.5 million people worldwide die prematurely every year due to air pollution and this comes from breathing in tiny liquid or solid particles that increase the risk of respiratory complications, strokes, heart disease, and sometimes even cancer. Air purifiers are intended to be a solution to this precise problem. Air filters are meant to improve the quality of the air and eliminate harmful substances through a variety of viable method mechanical filtering ionization and chemical methods (Xu, 2018)

4.2. THEORIES OF AIR FILTRATION TECHNOLOGY

The classic air filtration principles were first observed by botanist Brown in the early 19th century where he realized very fine particulates that were suspended in liquids and that defined the Brownian motion. The Brownian motion or pedesis is the simply erratic and random motion of microscopic particles suspended in a fluid as a result of continuous bombardment in the surrounding medium. It brought about the understanding of the law of aerosol filtration where these microscopic particles were discovered to have a maximum permeability within the radius of 0.1 to 0.2 micro millimeters. This was done by Freundlich in 1922 but the first to apply the pedesis theory was Kaufmann in 1936 who deduced the mathematical formula for a filtering effect from the fiber filtration theory. The idea here was simply recognizing that filtration was mainly affected by the interception of the particles and diffusion. This idea continues to be propagated and in 1967 the Pickaar and Clarenburg proposed a mathematical theory for fiber filter microporous structures using these classical theories but improving on them (Xu, 2018).

In recent years they have been theoretical simulations conducted by researchers on the formation of these microporous filters and mechanisms of filter cakes with some achievements. Without becoming entirely too scientific the inertia my capture mechanism and the way the air enters the filtration system are closely related to the mass of the particles and the overall efficiency of the air purification system. This means things like wind speed and the angle at which these particulate matters meet the fiber surface determines how the capture efficiency is influenced (Xu, 2013).

There are electrostatic capture mechanisms that improve the collection efficiency by creating an electric field using oppositely charged materials. These filter systems are produced by melt spinning a continuous filament which is a manufactured fiber (unless it is raw silk) that provides a combination of static and a structure which is able to filter efficiently. Is of shown that this continuous filament melts spun materials provide better initial and long-term filtration efficiency than the classic cotton or polyester medium. The employ triboelectrically charged materials which are mediums that become electrically charged when they are separated from other different materials they were initially in contact with. Examples include needled punched nonwoven technology used on wool and polypropylene fibers. The filtration efficiency of the quasi-permanent electrically charged materials was remarkably higher than the uncharged mediums and it is also important to note that the magnitude and duration of the electric charge are lower the bigger the fiber diameter is for the longest time Air filtration studies were focused on the collection efficiency of particulate matter and the drop in pressure. This becomes limited insignificant especially in high-efficiency low particulate air filters and the perspective changed into studying unsteady filtration processes that in the long run lead to increased operating costs and general inefficiency (Huber, 2004).

4.3. HIGH-EFFICIENCY PARTICULATE AIR FILTER SYSTEMS (HEPA)

High-efficiency particulate air absorbing is a high-efficiency standard air filtration system that was first invented in the US during the Manhattan

Project to control Radioactive contaminants that are in the air.(Whitt, 2020) Available at: https://www.nuaire.com/media/wysiwyg/pdf/use_of_hepa_filters.pdf It is most likely the biggest sleep in air cleaning technology and it was made from the aforementioned Brief History of air purification systems and although primarily used within the 1940s in a purely militaristic fashion in protecting soldiers and scientists from radioactive particulates. They became commercially used systems in the next decade Dennis way into the manufacturing industry what's contamination control from the production of disk drives medical devices semiconductors and pharmaceutical products. Since then it has made its way into our homes and vehicles as well as in hospitals and was capable of removing microscopic harmful air particles like pollen, pet dander, mold, and various other airborne pollutants. It is important firstly for us to understand how these particular materials compare in size and it is shown in Table 4.1.

Table 4.1: Particulates in the Air and Their Size in a Millionth of a Meter (a micro millimeter or micron)

Particle	Particle Size (microns)	Particle	Particle Size (microns)
Atmospheric Dust	0.001–40	Grain Dusts	5–1000
Bacteria	0.3–60	Household dust	0.05–100
Beach Sand	100–10000	Human Hair	40–300
Burning Wood	0.2–3	Insecticide Dusts	0.5–10
Cement Dust	3–100	Lead Dust	0.1–0.7
Clay, fine	0.5–1	Mold Spores	10–30
Coal Dust	1–100	Pet Dander	0.5–100
Combustion	0.01–0.1	Pollen	10–1000
Dust Mites	100–300	Smoke	0.01–0.1
Fly Ash	1–1000	Tobacco Smoke	0.01–4
		Viruses	0.005–0.3

Source: https://breathequality.com/hepa-filter/.

Within the last decade efficiency of single high fiber electrostatic dust filtration systems is actually expressed as logarithmic functions based on cyber services and the and volumes of particulate matter. This enables filtration systems of efficiency up to 99.9% especially when their filtration materials are taken into account. The Nano titanium dioxide photocatalytic material it's probably the most promising filtration system of the 21st century capable of eliminating harmful gases like various nitrous oxides sulfur dioxide and even ammonia.

So far we have seen HEPA () filtration systems are capable of an enhanced mechanism of diffusion which details the blockage of particulate matter through the Brownian motion, interception where particles flowing in the airstream adhere to the fibers especially those with very low radius and impaction where the inertia the particulate matter plays a role. As much as these processes are very efficient to arresting very fine particles they are not capable of filtering out gases and odors at the molecular level. This is where it is important to mention the above Nano titanium dioxide poly catalytic materials and other carbon cloth filters that are capable to go as far as gas filtration where it is important to separate volatile organic compounds chemical vapors flatulent orders and some use activated carbon in addition to the regular HEPA filter. These air purification systems go to the length of absorbing gaseous pollutants and a known as HAGA filters (high-efficiency gas absorption) and were initially adopted by the British military for the defense against chemical Warfare (Lafond, 2020).

HEPA filters sometimes coming conjunction with a pre-filter that is some form of activated carbon that can extend the lifespan of the costlier HEPA filter. This allows the larger particulate matter like dust and air particles or pollen particles before the second stage high-quality HEPA filter arrest the smaller and finer particles. There are six types of HEPA filtration systems as stipulated by the Institute of Environmental Science and Technology and each has its own distinct characteristics as shown in the chart table 4.2.

Table 4.2: Showing Categories of HEPA Filtration Systems According to the Institute of Environmental Science and Technology, (IEST)

Filter type	Penetration test		Scan test		Comments	Min efficiency rating
A	MIL-STD 282	Thermal DOP	None	None		99.97% at 0.3μm
B	MIL-STD 282	Thermal DOP	None	None	Two flow leak test	99.97% at 0.3μm
C	MIL-STD 282	Thermal DOP	Protometer	Polydisperse DOP		99.99% at 0.3μm
D	MIL-STD 282	Thermal DOP	Protometer	Polydisperse DOP		99.999% at 0.3μm
E	MIL-STD 51477	Thermal DOP	Protometer	Polydisperse DOP	Two flow leak test	99.97% at 0.3μm

F	IES-RP CC007	Open	Particle Counter	Open		99.999% at 0.1 to 0.2μm

Source: https://www.nafahq.org/how-many-types-of-hepa-filters-are-there/.

The overall specifications and standards adopted are dependent in which country and organization stipulate. In the US HEPA filters are defined by the United States Department of Energy and in Europe, specifications are determined by the European Union. Many companies may market their air filter systems using a term known as true HEPA to give their customers assurance however they still need to be tested in laboratories because some come reasonably close to the above-stated helper standards while others fall short significantly. True HEPA systems need to remove 99.97% of all particles of a size of 0.3 micro millimeters or greater. Testing is done in-place as well with the use of heterogenous dioctyl phthalate (DOP) or suitable substitutes to check for leaks and proper functioning. It also must be noted that modern HEPA filters are also fragile in nature and can easily be damaged during transit, if they are stacked inappropriately in trucks or punctured during handling or even during installation by incompetent technicians. A HEPA filter is only going to function optimally when gasketing and framing done well and tested during installation (Huber, 2004).

Several companies like Incen AG, IQAir North America, Austin Air Systems and Allen air have all done their part in the air filtration and air purification market by capitalizing in the technologies as they come. IQAir for example introduced a special air cleaning system for the 1990 Mercedes Benz model and later the H13 HEPA that traps 100 more pollutants than previous technology. Austin Air Systems improved the technology so that environment particulate contamination, unhealthy odors and chemical toxicity was more efficiently purified from air intakes by introducing the aforementioned pre-filter among other things.

With strengths in removing all sorts of dust mites and common Elder allergens together with not producing ozone

Or any harmful by-products HEPA air filter systems do not release the contaminants back into the air however this may count as a weakness because bacteria and viruses trapped in the filter can breed and reproduce. Complementary technologies like germicidal UV lamps can be placed in conjunction with the HEPA filter systems to stop microorganisms from reproducing (Lafond, 2020).

Figure 4.3: European Standards on HEPA Filters According to Their Designated Class

HEPA class	Efficiency
E10/H10	85%
E11/H11	95%
E12/H12	99.5%
H13	99.97%
H14	99.975%
U15	99.9975%
U16	99.99975%
U17	99.9999%

Source: https://breathequality.com/hepa-filter/.

4.4. ACTIVATED CARBON AIR FILTERS

Activated carbon air filters are especially good at getting rid of strong odors and other harmful gases like volatile organic compounds and gaseous pollutants from the air. Activated carbon air filters accomplish this in a different fashion from other air filters like HEPA that only filter particle pollution by using carbon Air filters that trap gases on a bed of charcoal. Typically, the pieces of carbon are in granular or powdered block forms and are treated to be very porous. You may find that one gram of activated carbon can have a surface area of 500 m squared which is very surprising but capable of absorbing more contaminants and pollutants. The activated carbon is chemically treated so that organic compounds cause them to stick and to them and the more porous the carbon is the more particulates it will capture.

Carbon purification has a colorful history perhaps thousands of years old wear charcoal has been used to purify water however to activate carbon that means processing it in a way that it is better at rapping gas molecules by first injecting it with hot air carbon dioxide or steam. What this does is allow a lattice of tiny pores within the carbon vastly increasing is surface area as stated above. The Egyptians are perhaps the earliest recorded people to have used carbon to remove odors from wounds that are infected. It has also been recorded that sailors have used charcoal to start drinking water in barrels. They would line the inside of these barrels with charcoal to keep water fresh on their long expeditions. World War 1 also presented charcoal filtration systems in gas masks as it was effective against deadly gases and toxins. This was perpetuated further in World War II and that's led to the development of the modern versions of air purification systems as well as water filters (Huber, 2004).

Activated carbon air filtration systems used a process called adsorption to make particulates adhere to the outside of carbon surfaces. Note this is different from absorption where there is a chemical bond at a molecular level. Once this filter bed is filled up it can no longer trap anymore pollutants and must be replaced. The adsorption qualities of activated carbon systems are dependent on temperature and relative humidity as well as and its ability to be bonded chemically due to different levels of activation (Figure 4.1).

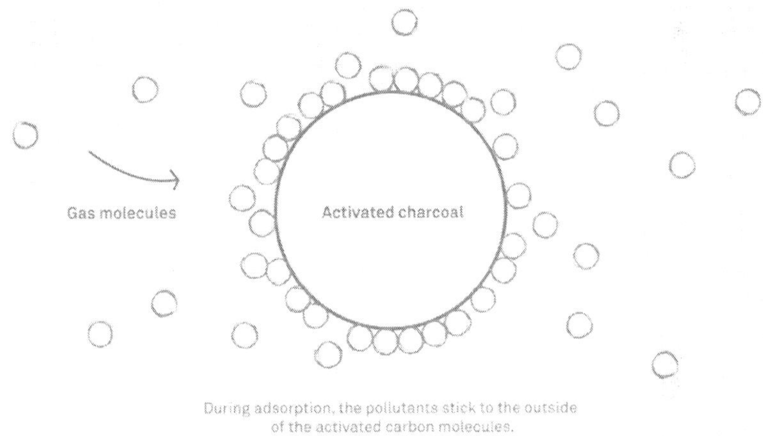

Figure 4.1: Diagram of how carbon air filters trap gaseous pollutants.

Source: https://molekule.science/activated-carbon-air-filter/.

Activated carbon filters can be used alone but more often than not they are used in conjunction with other filters especially the high-efficiency particulate air filters. They are both in home office settings that are small and in larger industrial settings because they are low cost and have a myriad of uses from industrial application to medical and environmental agricultural uses. It is important to know how to effectively use your cabin air filter and this requires you to make sure that the airflow is adequate enough to remove most of the pollutants by maximizing the amount of time spent by the air passing to the activated carbon. In the industry this is known as dwell time and allows the filter to remove most of the pollutants possible. Enough carbon should also be present in the system as some filters as much as they claim to be activated carbon only use just a trace of and can become saturated fairly quickly (Xu, 2013) (Figure 4.2).

Figure 4.2: A micrograph scan of activated charcoal under electron microscope.

Source: https://en.wikipedia.org/wiki/Activated_carbon#Classification.

The technology was highlighted in 2015 when absorption-based granular activated carbon filters were experimented on by Khazraei Vizhemehr and his associates. We discovered that it was very efficient in removing toluene both in single and mixed scenarios with other air purification systems. Among other physical properties of indoor contaminants, it was found that molecular weight is what will determine the performance and service life of tested granular activated carbon filters (Xu, 2018)

Once a carbon filter is saturated it loses its effectiveness and must be replaced. One way this can be easily noted is by the increase in the smell or odor that you may notice however manufacturers usually have a guideline on how regularly the change should be made or at least just a rough estimate. This may actually be the only drawback of using activated carbon systems as well as the fact that it does not remove many particulate pollutants that do not adsorb to the carbon. Both HEPA filter systems and activated carbon systems require a fan to draw in air in order to trap the particulates and in this way they are said to operate passively. They also do not effectively capture bacteria and viruses it may in turn provide a breeding ground for them as the pores in the filter become trapped and effectiveness is lowered. This is known as filtration colonization. Fungi may grow and utilize these areas of organic particles as a food source as their captain and then spread downstream as air pollution. Technologies to develop antimicrobial properties on the filters are still underway especially in conjunction with Nanofibers and nylon (Xu, 2018) (Figure 4.3).

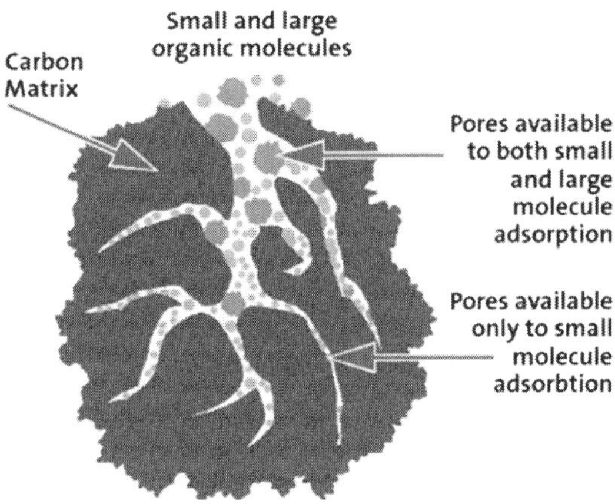

Figure 4.3: Adsorption of particulates onto activated carbon.

Source: https://cdn-foobot.pressidium.com/wp-content/uploads/2018/06/Activated-Carbon-300x245.jpg.

4.5. AIR IONIZERS

Air ionizers as the name dictate ionize electrical charging air molecules and in this way purify. A negative ion generator is a device that uses High Voltage and electrically charged molecules purifying them. An ion essentially is a particle or an atom that is charged electrically and negative ions have one more extra electron that is also known as anions. Cations are positive ions or in other words, particles are atoms that have a missing electron that result in a net positive charge. The technology is accredited to Cecil Alfred laws nicknamed copy was a British electronic engineer during the Second World

War and his invention is accredited to having the capability of eliminating even bacterial infections and lower airborne viral infections buy statically charging ambient air. Instead of the aforementioned HEPA filters and carbon activated air purifiers ionic filters do not have to use a fan to pull in polluted air but produce a cloud of negative ions to electrically charged molecules and attract airborne particles. After this, they fall onto a charge collector plate or mechanical air filter for the need to get earthed or grounded. It must be noted that not all ionic air purifiers are fanless but the ones that are a bit noisier and use a lot more energy circulating air around the room (Lafond, 2020).

The National Health Service carried out research on the effectiveness of onions for air purification because it was a high frequency of nosocomial infections in the hospitals. Infection rates for a certain bacteria known as a senior tube actor that is negatively charged with dropped 20 because of the effectiveness of the negative air ionizer and further research has been carried out on various other bacteria and viruses it has been found that positive and negative ions produced by air ionizers render many of them inactive including influenza. The negatively charged ions attach two positively charged particles in the microbe: the result are bacteria or allergens that cannot float in the air and must seek a method to ground themselves electrically (Lafond, 2020).

This prompted further research was carried out to find out if ionizers are good for people's health and evidence shows that the significant impact it had on people suffering from seasonal allergies and even asthma sufferers noticed a difference in the quality of the air means that it did a lot more good than harm. The answer however was still is not a resounding yes or no. This is because as much as ionizers use electrostatically charged plates and produce both negatively and positively charged gas is even the best of them produce a small amount of Ozone which is a triatomic oxygen Bond that is dangerous to one's health. Ozone from your air ionizer can irritate asthma sufferers and exacerbate many other breathing issues. Air ionizers also Safari's downfall of not being able to remove orders on smells and generally being more power-consuming than other air filtration systems. They do have however a great performance of removing bacteria and viruses and some like the Air Oasis air ionizers claim that the kill microbes like most protozoa fungi and mold. Table 4.3 shows the difference between a classic air purifier and a pure air ionizer (Lafond, 2020).

Table 4.3: Comparison of Air Purifier vs Air Ionizer

	Classic air purifiers	**Air purifiers with ionic feature**	**Air ionizers**
Fan	Yes	Yes	No
Annual filter replacement	Yes	Maybe	No
Power Consumption	Normal	Higher	Very high
Sound	Silent at low speed	Highest	Inevitable noise
Performance	High	High	Random

Source: https://breathequality.com/ionizer/.

4.6. ELECTROSTATIC PRECIPITATORS

Electrostatic precipitators induce an electrostatic charge on air that is flowing into the machine and filters out fine particles like dust and smoke by impeding of the flow gases. Electrostatic precipitator supplies the electric charge on the object with water that is being collected and in essence, this means that energy consumption is low and the Machine overall has great efficiency. Its history goes as far back as 1824 where a Scientist called Hohfeild used an electrostatic Corona discharge precipitator to remove particles from an aerosol and this was only done within laboratory settings and commercialized much later. A corona discharge is not like the name is sound but is an electrical discharge what about when a fluid has ionized the conductor or in this case electrodes that really carrying a static charge. It is however an unwanted side effect as Corona is an unwanted side effect and is economically a significant waste of energy because of the high voltage output and care must be taken to live in the electrical field strength within the whole system. This can be done by insulating Corona rings and that is why you will see electrostatic precipitators having a series of rubes in which electrodes are fitted (Xu, 2013).

Corona discharges have a variety of commercial and industrial uses and are fitted into electrostatic precipitators to remove solid particulates from customs as well as in air ionizers as far as air conditioning and air purification systems go (Figure 4.4).

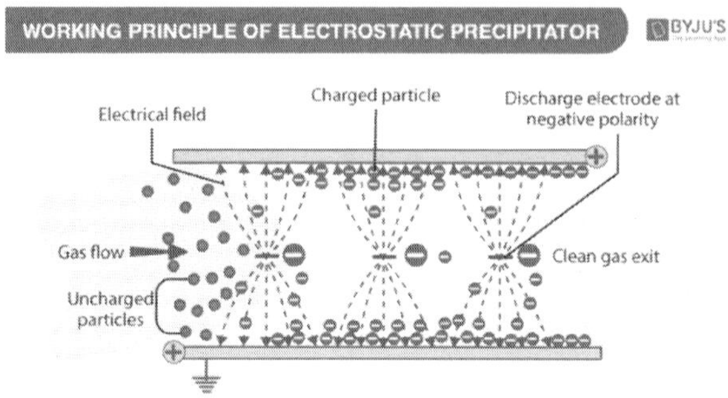

Figure 4.4: Working principle of the electrostatic precipitator.

Source: https://byjus.com/physics/electrostatic-precipitator/.

4.6.1. Different Types of Electrostatic Precipitators

- **Plate Precipitators:** delete all the classic potato types and they are built up of a series of things vertical wires and arrange next two large flat metal plates usually a distance of a little was 1 cm to 18 cm on large-scale electrostatic precipitators. The airflow passes horizontally through the plates and since the air between the vertical wires and the plates is negatively charged, particles are attracted to the plates via electrostatic force to get grounded. The collection plate is then emptied.
- **Dry Electrostatic Precipitators**: dry electrostatic precipitators are ordinarily used to collect particulates like ash or cement in a dry state. The air flows through a Hooper where the particulates are collected and removed by hammering the electrodes.
- **Wet Electrostatic Precipitators**: these are precipitators that intended to remove resin oil or tar from the intended fluids and the collectors need to be continuously sprayed with water to remove the unwanted materials. They are built to be more efficient than dry electrostatic precipitators.
- **Tubular Precipitators**: these are potatoes that are a single unit consisting of tubes where High Voltage electrodes are placed parallel to each other and the arrangement can you be circular of hexagonal like honeycombs with gas flowing through the tubes and particulate matter is collected.

Electrostatic precipitators are built to be of high durability and can collect both dry and wet impurities with the most modern ones having low operation costs. The physics and chemistry around the inner workings of electrostatic precipitators make it very efficient in collecting even the smallest particles while having large gas volumes even at the lowest pressures with High dust loads. They however cannot be used for gaseous emissions and consume a lot of space. Although the application is wide they are not adaptive to change an operating conditions and the initial capital investment is quite high (Lafond, 2020).

Electrostatic capture mechanisms also consist of the polarized Media electronic air cleaners that use electricity as well as passive mechanical filters. It is known that most airborne particles have a charge and many are even bipolar. As particles pass through electric fields they adhere to fiber Media pads where they can be easily disposed of. Ultrafine particles that are not collected on the initial pass through the system have to be grounded because they are charged and so agglomerate to other particles and then are later collected through subsequent passes. It must be noted that this is a non-ionizing technology that is not likely to produce any ozone (Lafond, 2020).

4.7. OZONE GENERATORS

Ozone generators charge air particles and deliberately produce ozone as a byproduct which is a strong oxidant gas and in turn oxidizes many other chemicals. This technology is only safe in rooms that are not occupied and commercial ozone generators can produce up to 3000 milligrams of Ozone per hour. This is generally known as shock treatment for a room that removes smoke orders after fires and musty smells produced after flood damage including mold from walls and floors some that may be toxic. The smell that is normally produced by decaying flesh that sometimes cannot be removed by bleach and other chemicals will ordinarily be removed with the use of ozone and although it is not a healthy gas it can be useful. Lafond (2020) noted that ozone is a highly toxic extremely reactive gas that damages the lungs and of faculty about cells directly and any air purifier that has an output of more than 0.1 ppm.

It essentially works using the same technology as electrostatic precipitators using a silent Corona discharge method and splitting oxygen molecules into a single atom where do letters to other oxygen molecules to create ozone. As mentioned previously ozone has irreversible health effects and the room needs to be on occupied for at least several hours until the ozone clears off. There are ozonators that also produce ozone using ultraviolet (UV) light (Lafond, 2020).

4.8. PHOTOCATALYTIC OXIDATION AIR PURIFIERS

Photocatalytic air purifiers can improve overall air quality by using a specialized air filtration device that accelerates photo reactions which are chemical reactions what are concerned with lights with the use of a catalyst. They are advanced technology and fairly new to the industry based on research from NASA and the United Technologies Research Center as a compact efficient filter. Give him water resources and as much as it's a generic term it is also chemical in nature with an activated catalyst where chemical changes usually oxidation and in some cases a metal oxide cause semiconductor usually titanium allow absorption of a UV photon to make chemical changes (Xu, 2018).

The history of photocatalysts research has been a long but difficult Road mainly due to lack of interest and lack of funding but it began in 1911 with Dr Alexander Eiberg illuminating zinc oxide. Breakthroughs in research did not occur until Akira Fujishima and Katy Honda discovered electrochemical fertilizers of water occurring between titanium Oxide and platinum electrodes. This was the first time that hydrogen production was produced from a clean and cost-effective source. Although research has been on and off the most applicable in relation to air purification technologies has been the Light 2CAT project by the European Union where light-sensitive materials are being developed and incorporated into structures that combat air pollution.

Photocatalytic oxidation is a continuing technology that means a lot of improvement and especially when it comes to the UV sterilization bulbs but needs to be replaced at least once a year and the whole system in general orphan comes at a high cost. Featured technology is the photoelectrochemical oxidation of purification that involves electrochemical interactions together with the catalyst material to produce clean air (Xu, 2018).

Another type of catalytic filter is the cold catalytic filter that works by using a low-temperature and low-adsorption catalyst that rapidly decomposes and breaks down particulates more specifically organic compounds in the air with oxygen to produce water and other substances that are harmless to the human body. Also known as the low-temperature catalytic filter, they are capable of decomposing formaldehyde and dust particles in the air while also removing other harmful gases such as benzene, ammonia, and other volatile organic compounds. They do not require UV fixtures and therefore have a longer lifespan than their cousin the photocatalytic filters.

4.9. UV AIR PURIFIERS

UV air filters are Common purification technology used in many households that are certified by many different agencies and are deemed 100% safe although it is a form of radiation and it should never be pointed directly towards a human being. Light is made up of particles known as photons and UV light is made of the shorter wavelength of life then what is visible to the human light about 100 to 400 NM They are used even by the center for disease control because they're capable of rendering pathogens viruses and harmful germs in Earth because they disrupt these pathogens at the core DNA and prevents them from multiplying. Short exposure to this light trend is he surfaced disinfected from any infectious disease. Officially we can call it UV germicidal irradiation and the use of UV lamps via a forced air method. Once installed the air circulates in the room and is forced through the system and microorganisms past of the Lambs and sterilized. It has been an accepted practice to use UV light for sanitation purposes since the mid-twentieth century and can even be used to sterilize drinking water and even and wastewater. The effectiveness is dependent on the length of time that the ambient air is exposed to UV radiation and the intensity of the wavelength of light (Huber, 2004) (Figure 4.5).

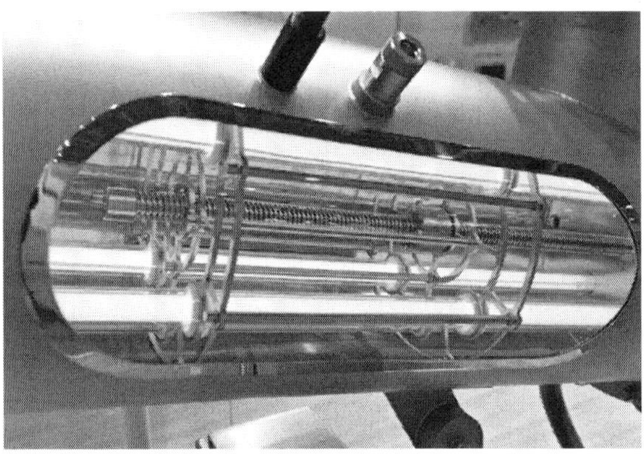

Figure 4.5: UV light air purifiers.

Source: https://molekule.science/how-do-uv-light-air-purifiers-work/.

Safety must be taken when using UV systems as skin exposure can result in rapid sunburn and skin cancer and exposure to the eyes can be extremely painful and can cause permanent vision impairment and inflammation of the cornea and blindness and some cases. UV light also produces ozone as a reaction to oxygen molecules and it is harmful to one's health. UVC C (100–280 nm) that is responsible for that disinfectant activity has a lot more extra energy behind more than the visible blue light that you ordinarily see and DNA is particularly susceptible to UV light bombardments. UVC air filtration systems also do not remove borders unless they are designed to produce ozone as well and it has no effect on particulates like pollen and dust including most allergens as much as it is effective against microorganisms (Huber, 2004).

4.10. THERMODYNAMIC STERILIZATION

Thermodynamic sterilization is a technology that uses heat to treat us more specifically Eros transferred buy air convection to a ceramic coil with microcapillaries which a heated in excess of 200 degrees Celsius and incinerates 99.9%of microbiological particles. Like photocatalytic oxidation it is a fairly new advancement in technology and is expensive and extremely energy-consuming. Once the bacteria viruses mold and fungus spores are incinerated by the ceramic core the air has to be cooled through heat transfer plates as it is then released. The biggest drawback with this technology although it is still in the theoretical phases is the byproducts of partial thermal decomposition particulate which are still not yet addressed. The proponents of thermodynamic sterilization technology claim that they do not emit anything harmful and on the contrary actually reduce the concentration of Ozone in the atmosphere.

4.11. IMMOBILIZED CELL TECHNOLOGY

Immobilized cell technology uses the same principle as air ionizers in the sense that it depends on particulate matter and the child that they carry either negative or positive. It is a method of air filtration and purification that uses whole-cell immobilization which is an enzymatic process where microfine particulate matter is separated from the air by attracting charged particles with the use of a bioreactor to enzymatically render them inert. A bioreactor is a vessel that through a chemical process changes an organism's biochemical structure either aerobically or aerobically.

The system works by moving air into the cell technology system and the particles that are charged fall onto a water cascade and then the particulate matter is pulled into the bioreactor. The bioreactor breakdowns all these particulate matters into their base elements and the immobilized cell technology is noted to be 12 times more efficient than natural oxidation without the employment of mesh filters and purely through water, electrical power, and the bioreactor. The base elements that are inert need to be removed periodically.

Listed above are all the encompassing air purifying techniques and technologies both active and passive. The active air purifiers change the ions in the air causing pollutants to stick to whatever surfaces while the passive air purification units are the above that use air filters to catch and remove pollutants. As it can be used to deduce the passive purifiers are a bit more efficient since all the particulate matter is permanently removed from the bio reactor and collected into filters. The ones that are most commonly used are the HEPA filters and germicidal irradiation or UV air purification system but what is important to note is that many filtration techniques can be used in combination as the diagram below dictates (Sutherland, 2004) (Figure 4.6).

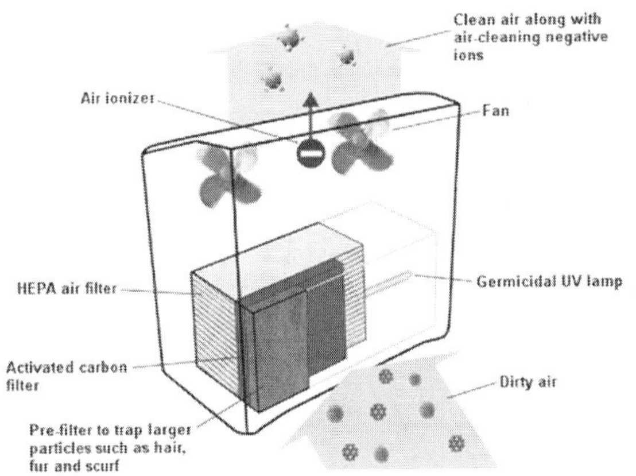

Figure 4.6: A sample air purifier that incorporates multiple technologies.

Source: https://blissair.com/air-purifying-technology-overview.htm#carbon.

The market today has seen high efficiency particulate arrestance filters polyester filters are repeated and fiberglass filters in use in homes and workplaces each have their own benefits and drawbacks and one must choose the right filter for air quality needs. (Engineering Toolbox, 2005) Available at: https://www.engineeringtoolbox.com/filters-efficiency-arrestance-d_958.html

4.12. PLANTS AND TREES

Plants and trees are the most natural and most overlooked purifying devices and is what ordinarily occurs in nature. Keeping house plant or trees inside your home improves the quality of the air drastically and new blue effects a better psychologically. Naturally, plants and trees are designed to purify air for humans and consume carbon dioxide that human beings and animals expel turning it into pure fresh oxygen which anyone who did Elementary science would attest to. The overall task is to keep the plants alive which is not difficult at all and some may even enjoy it. Within the whole list of air purifying technology within the whole list of air purifying technology this is probably the cheapest (OECD, 2011).

The most effective fashion this has been done is by creating active living walls that act as passive filtration systems where you allow the development of vegetation along vertical surfaces not even go so far as to create microclimates and is known to be the most easily adopted to mechanically ventilated buildings. What this means is that it can be integrated into the building's air-conditioning and ventilation Systems cooling purifying and humidifying the air.

They are a great alternative to energy-consuming artificial air purifying systems and results show that they remove significant amounts of air pollutants from 80% of formaldehyde 50% toluene and 10% trichloroethylene reductions recorded. This has been adopted into integrating air filters and purification technologies with building information modeling platforms that have attractive a good amount of stakeholders (OECD, 2011).

4.13. AIR FILTRATION MATERIALS

Different materials have been deployed and developed overtime to filter air and improve ambient air quality. Gas fiber air filter materials for example appeared in the US in the 1940s and have continuously been developed and fit in many HEPA filters with super fine glass fiber paper that can achieve 99.9 998% efficiency for particular and dust that are less than 0.3 micrometer. Activated carbon has been propagated into activated carbon fiber sheets developed by the Japanese in the 70s with rayon and properly nitrile and other raw materials and is considered to be one of the best air purification systems in the 21st century because of its enormous absorption abilities. Even our new fiber that was a technology born in the late 80s has made its way into the air filtration industry because of its largest specific surface area surface energy and surface tension that increases efficiency in collecting airborne particulates (Sutherland, 2004) (Figure 4.7).

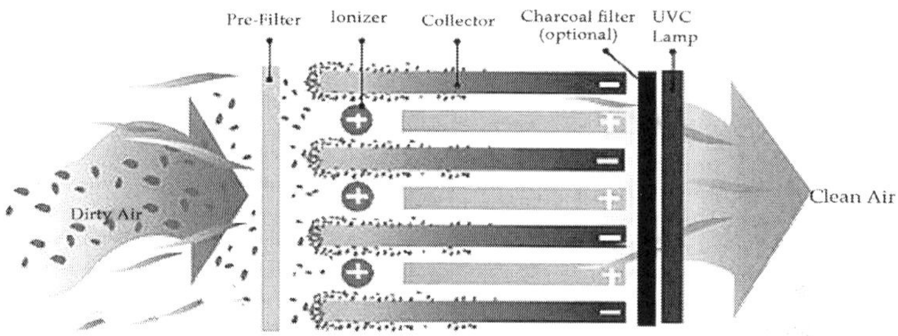

Figure 4.7: A sample air purifier that incorporates multiple technologies in this case an electrostatic filter with activated charcoal and a UV lamp.

Source: https://blissair.com/air-purifying-technology-overview.htm#carbon.

Companies like Lydair in the United States were able to produce ultra-low penetration air filters (ULPA) that went as far as cleaning the air with dust particulates of 0.1 micro millimeters proficiency levels of 99.999% and helped develop precision machinery used in the aerospace and electronics Industries. They developed the film compound filter material from polytetrafluoroethylene and covered in a fabric of non-woven cloth in the late 1990s and it quickly became popular with purification and air filtration Industries. It proved to be durable and had a long service life of the five times the traditional filter material and could withstand temperatures of up to 280 degrees Celsius.

Nano titanium dioxide photocatalytic material that has been mentioned above but needs to be retreated as one of the most promising filtration materials of the 21st century using UV excitation energy leading to oxidation reduction reactions and works on other corresponding inorganic material as well. In conditions where traditional filter materials would have low filtering efficiency or high resistance, electric filter materials and nano titanium dioxide has proved to be more effective and has been able to garner more public attention due to its low pressure drop and better filtration performance.

4.14. AMBIENT AND INDOOR AIR QUALITY STANDARDS

The United States of America was the first to formulate national ambient air quality standards (NAAQS) followed by the Australian government, the European Commission, the Japanese government, and the WHO. General ranking is based on particulate matter and a performance index with efficiency under a working velocity. The Australians were the first to meets WHO standards and surpass them with Japan and the USA closely behind and currently the Chinese appeared to have a long way to go in relation to environmental air pollution and air quality standard (Oppenländer, 2007)

Indoor air quality standards for particularate matters varies country-by-country with only a few countries able to establish the lowest control standards and recommended limits but the important thing to gather from all of this is that the air filter standards need to be a lot more stringent as well to ensure that the quality of air is healthy to breathe. The test methods had to be developed and continuously monitored. For the most part it has been done very well starting with the USA in 1938 and many European countries following in the 50s and 60s. In the 80s the Chinese also proposed atmospheric particulate measures and air filter standards of classification.

The measurement of pollutants does not differ wildly on a global scale and main pollutants in the air can be divided into three categories suspended particles volatile organic pollutants and microorganisms. Suspended air particles can be removed from the Earth with the use of purification Technologies like filtration electrostatic precipitation and onion technology. Filtration is still the most widely used to fit Asian system for suspended particles and its simplicity low cost and effectiveness continues to ensure this. Activated carbon activated carbon is also effective for suspended particles to a certain extent and is most used for absorbing volatile organic pollutants. In addition, photo catalytic and plasma cleaning technologies are used effectively against volatile organic pollutants to purify the air. Against microorganisms we have already discussed the efficiency that UV light has especially in conjunction with photocatalytic technology although bacteria with larger diameter can be stopped even with some effective filters while they may not be suitable for the elimination of viruses (Huber, 2004) (Table 4.4)

Table 4.4: Effect of Single Purification Technology to Main Types of Pollutants

Purification Technique	(Particulates per category)			
	(Suspended particles)	(Volatility organic contaminants)	(Microorganisms)	
	(Dust, pollen, secondary pollutants, lampblack, etc.)	(Formaldehyde, benzene, ammonia, etc.)	(Bacteria)	(Virus)
Particulate size	Diameter 0.01–100 μm	Diameter 0.0001–0. 001 μm	Diameter 0.2–10 μm	Diameter 0.01–0.3 μm
Filtration	Effective	Not effective	Effective	Not effective
Adsorption	Partially effective	High-efficiency	Partially effective	Not effective
Water washing purification	Effective	Partially effective	Not effective	Not effective
Electrostatic precipitation	Effective	Not obvious	Partially effective	Not effective
Anion technology	Effective	Not obvious	Partially effective	Not effective
Photocatalysis purifying technology	Not obvious	Effective	Effective	Effective
Plasma cleaning technology	Not obvious	Effective	Effective	Effective
Ultraviolet radiation	Not effective	Not effective	High-efficiency	High-efficiency

Source: https://www.sciencedirect.com/science/article/abs/pii/S221067071630734X.

This is the reason that ambient air, indoor or out especially with all its multiplicity of air pollutants cannot use a single technology to purify the air successfully and needs many of them in conjunction in order to cater for the disadvantages where they may be any. Indoor air pollution purification Technologies were developed initially with the intention of treating industrial exhaust gases and other gas treatment Technologies. This is the reason that a synergistic effort was brought about with integration of other purification Technologies (Huber, 2004).

Most air filter it is in purifiers in the market today adopt HEPA filters although they may have a pre-filter medium filter or ultra-low particulate air filters. Ultra-low particular air filters catch 99.999% of particulate matter 0.122 0.17 micro millimeters in diameter however HEPA filters are designed to be over 99.99% efficient as well and are used in theatres hospital respirators and even vehicles. Water filtration techniques include the Trombe wall that has high air filtration efficiency together with a biofiltration technique because the trump wall is ineffective against particles of 0.1 millimeters in diameter. The procedure begins with air that is polluted going through the biofilter where the contaminants are degraded buy micro biotic oxidation essentially in the same principle as an immobilized cell air filter eliminating volatile organic compounds (Huber, 2004) (Figure 4.8).

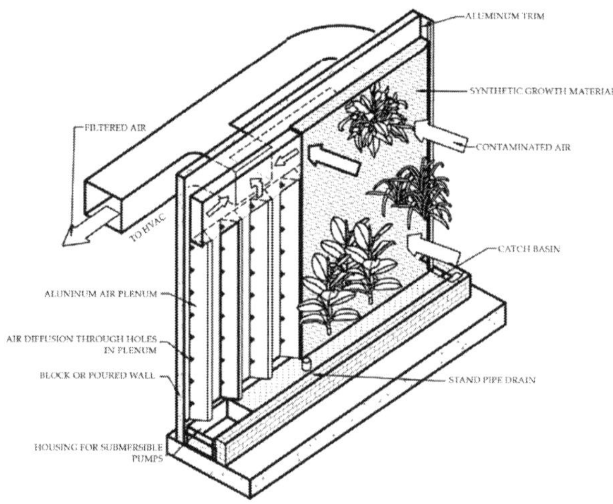

Figure 4.8: Schematic of indoor biofiltration system.

Source: https://www.sciencedirect.com/science/article/abs/pii/S221067071630734X.

4.15. ENERGY ASSESSMENTS AND OTHER CONSIDERATIONS OF AIR FILTRATION AND PURIFICATION SYSTEMS

It can be considered that every furcation system across the market offer similar performance however latest developments have shown that this is not actually the case and due to commodity status price has become a crucial factor and it is important for us to assess the reasons why. First of all, we must assume that these air filtration systems are intended to operate 24–7–365 and that accounts for 81% of the filters total life-cycle cost which many times can be exponential in relation to the art of the initial investment and maintenance cost that is a remaining 18%. The remaining 1% is to account for the disposal costs at the end of the air filtration and purification systems life cycle. In light of that information, we should consider that the total life-cycle cost of a particular filter or purification system that is energy efficient may initially cost less that's actually achieving energy conservation without much additional investment (Sutherland, 2004). When it comes to more active purification systems air pressure drop of a filter greatly affects the energy consumption overall and this is another very important factor to observe. Table 4.5 shows the specific parameters considering different domestic air purifiers please note that the average cost per year is in the Chinese Yuan.

Table 4.5: Specific Parameters of Different Domestic Air Purifiers

Origin	Brand	Reference Model	Applying	Rated	Noise level	Clean Air Delivery	Average Air-	HEPA *Service*	Average HEPA
			Space (m^2)	Power (W)	(dB)	Rate (CADR) (m^2/h)	flow rate (m/s)	Life (Year)	Cost (CNY/year)
Europe	Philips	AC4374	25-40	56	35-65	301-400	3.1-6.8	1	600
		AC6608	50-85	75	34 -65	400-710			
		AC8612	64-109	105	34-68	400-910			
	Blueair	303	17-30	13-60	34.7-55.9	105-205	1.3-4.6	0.5-1	296
		450E	31-50	30-80	32-52	401-500			
		Pro XL	56-96	30-120	39.5-64	800			
	Electrolux	EAC103	30-39	150	28.5-43.3	193.5	1.7-7.2	1-1.5	329
		EAC403	40-59	150	27-60	579			
		CN500AZ	40-59	125	28-51	439			
Japan	Sharp	FU-GFM50•13	30	55	65	312	3.1-6.3	5-10	122
		FX.CF100.N	80	8-118	38-68	800			
		KI-131360-W	37	62	62	369			
	Panasonic	F-VX670C-N	25-44	10-50	20-50	368	1.6-5.8	1	240
		F-VJL55C	31-52	10-53	22-53	322			
		F-VJL90C	100	11-88	23-55	501-700			
	Daikin	MCK38 RV2C-N	21-30	9-32	34-60	200-228	0.8-8.3	2	199
		TCK70P-T	50	12-75	23-54	401-500			
		MCKS7LMV2	21-34	8-39	31-60	270-342			
the United	Honeywell	KJ300E-PAC11016	21-36	7.5-55	65	305	3.3-6.1	0.5-1	219

States		KJ450E-JAC2022S	31-54	2-66	67.1	401-500			
		K1700E-PAC2127W	49-84	72	69	243-700			
China	Yadu	KJ500G-S5	35-60	80	s65	577	1.2-6	0.5-1	329
		KJG3066DR	> 60	60	s48	> 360			
		KJG588G-P5	80-100	80	565	588		1	
	Broad	TA240	12-20	10-30	20-45	240	-	0.25	198
		TA1000	41-60	13-40	25-45	1000			
		TA2000	> 61	30-110	30-53	2000			
	lexy	KJ703-F	51-99	55	32	100-550	3.4-8.1	0.5-1	249
		KJ601	51-99	60	66	301-400			
		KJ801	60-100	60	66	480-840			
	TCL	TKJ300E-S102	10-60	68	17-35	300	3.2-7.2	1	288
		F220B	30-60	45	s 35	220			
		TKJ400A5	40-60	45	20-40	403			
South Korea	Samsung	AX022FCVAUW/SC	530	10.8	30-40	179	2.2-5.4	2	300
		KJ393G-K5050WD	47	45	s61	393			
		KJ720E-K7586WF	50-86	100	29-67	719			

Source: https://www.sciencedirect.com/science/article/abs/pii/S221067071630734X.

The next consideration that should be taken into account when marketing air purification and filtration systems are thermal comforts. This is indicated by temperature but it is not the only factor because aspects of velocity and humidity are also significant in determining thermal comfort. A standard heated indoor environment can cause people to feel stuffy showing that average velocity for purifiers should be adjusted and normally range from 0.8 m per second to 8.30 per second. The way the air moves can result in exposed body parts having sensations of coldness and discomfort so higher temperatures are required some brands take this into account but others like the Chinese brand broad don't even mention air flow rate of the product specifications. The fresh and clean air that is dispersed by air purification systems also must be distributed evenly within the occupants breathing zones within an indoor environment. The recommended indoor humidity range for air conditioning is somewhere between the range of 30 and 60% depending on the season. Humidity levels are high making people uncomfortable and reducing it can make a person feel cool dry and comfortable. Unfortunately, many of these parameters are not considered the domestic air purifier or indoor air filter systems and we continue to need left the subject (Xu, 2018).

Other considerations include the acoustic impact that air purifiers have. Noise can have adverse health effects physiological as well as psychological and because of this many governments have established standard for controlling levels of machine noise. The Chinese dictate that national standards should be a clean air delivery rate (CADR) of 150 m cubed per hour or less than that while noise levels should be equal to or less than 55 decibels. For machines that deliver between 150-meter cube per hour and up to 300 meters cubed per hour of CADR needs to have a machine a produces less than 65 decibels worth of machine noise. These are just standards on the Chinese market however there is a lack of international noise standards because different countries adopt different noise levels and standards. The reason this exists is simply because of lack of detailed research.

Maintenance as an economic assessment is also neglected both to in the research process and overall during the lifetime of the air filtration of purification system. Filters need to be cleaned regularly or replaced periodically otherwise the system just becomes another source of pollution and it is obvious that the longer the filter works the more troublesome and like it is to be potentially polluting the air rather than cleaning it. HEPA air filters which are the most popular in the market today have a high maintenance cost because it needs a filter change once every year while some filters need to be changed even in as little as half a year (Xu, 2018)

In conclusion it is important to say that sustainable and healthy indoor environments and air purification and filtration technologies have received a good amount of attention over the last decade with studies aiming to reduce any possible source of air contaminants and particulates. As the traditional fibrous filters have various advantages they have being characterized to reduce pollution during high-pressure drops and the high maintenance costs have made them less attractive. The strategic combinations filtration techniques and other purification technologies have improved performance exponentially.

Electrostatic air filters together with the Trumbe wall system have shown how biofiltration and mechanical active purification techniques can be effective when applied in a synergistic effect. As much as many underlying factors like noise pollution have not been heavily researched it appears that direction is being taken especially taking into account filtration technologies that use building information modeling platforms like bio Facades and active living walls. Together with providing an outline of the technologies that exist today in the air filtration and purification industry this chapter also provides the guideline towards future air filtration technology research considering the point that have been lacking towards the agenda of achieving sustainable and healthy air and building ventilation systems.

Due to the nature of air purifier systems becoming more and more diverse it is a very relevant undertaking to look at the various air filter and purifying systems in the market today. Picking an air purifier for your home, office or vehicle may seem straightforward forward but as shown above, it is a lot trickier than it appears. The Alen Breathe Smart Classic Large Room Air Purifier although it may be slightly costlier on average has proved itself very efficient and popular in the market today, cleaning up to 1100 meters cubed of air in less than 30 minutes. Even on turbo mode (the highest setting), it has effective noise reduction technology that is very impressive. It is appropriate for large homes and comes with a sensor that indicates the real-time purity of the air in your large home.

Levoit Compact Air Purifier is another system that has made a big splash in the market most especially due to its high effectiveness (99.99%). It is comprised of a prefilter, a HEPA filter, and a granulated activated carbon filter with a three fan stage setting that is meant to control both the speed and volume of air purified. Ideally, it was intended to be for 1 room use, probably the bedroom, but its small size means you can carry it wherever you go. Honeywell True HEPA Allergen Air Purifier is a state of the art purifier that is meant specifically for the use of homeowners with pets. Allergists recommend it time and time again due to its effectiveness against pollen, pet dander, and general dust particulates. For its size, one will be very impressed that it can actually churn out fresh air up to 5 times in 1 hour. If one is seeking air purifies and filters it is not just this model but most products on the Honeywell line are designed to be more than suitable for those suffering from allergies. Hathaspace Smart True HEPA is an impressive air purifier with noise levels around 30 decibels which is the same as hearing leaves rustle in the wind, or a whisper. They were designed specifically to combat bacteria and dust and it delivers with its compact dimension of 27 inches in height and 16.5 inches in width. It has a complicated five-stage cleanup system consisting of:

- A pre-filter system designed to remove large particles like hair;
- A second pre-filter made of polypropylene that has an antibacterial coating, meant mainly to protect the HEPA filter;
- A HEPA filter with 99.99% effectiveness against particulate matter with a 0.3 micro millimeter diameter that would filter out dust, dust mites, mold spores, pet dander, pollen, and fungi;
- A granulated activated carbon filter that would offer the odor cleansing properties as already described above; and

- A Cold Catalyst filter is the last stage that targets formaldehyde, benzene, and other volatile organic compounds.

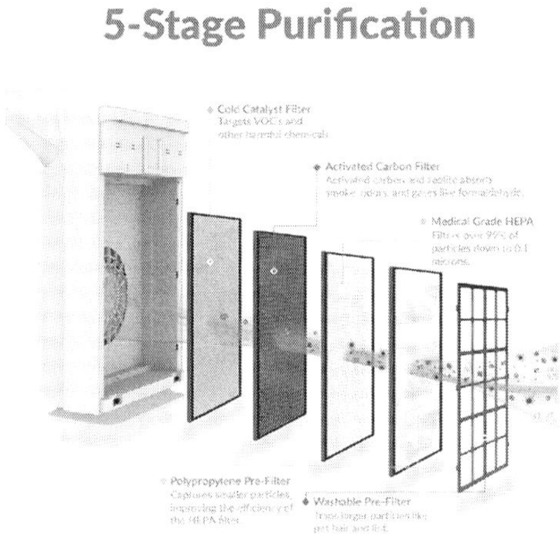

Figure 4.9: The five-stage purification system of the Hathaspace smart true HEPA.

Source: https://breathequality.com/hathaspace-air-purifer-2-review/.

When shopping for an air purifier, the finances are usually the first consideration but one must also appraise the CADR as already expressed. This rating that was created by the Association of Home Appliance Manufacturers (AHAM) effectively communicates the efficiency levels of an air purifier. The higher CADR, the better the air purifier is at providing fresh and clean air to the room. ACH is another consideration to take into account and it stands for air changes per hour. This communicates you how many times in a single hour that the purifier swaps out the air in the room. An air changes per hour rating of 5x means that the total volume of the air in the room will be exchanged five times per hour (Lafond, 2020) (Figure 4.9).

Another factor to consider is the energy star label of the air purification system. This trusted, government-backed, USEPA established label specifies household electronics that operate using less electricity. This offers more energy efficiency and savings on energy bills without sacrificing performance. Air Quality Sensors are another feature to consider and although Lower end models do not monitor the air quality, higher-end models tend to include air quality sensors that actually regulate when and how the system functions. They monitor the environment for particulate matter and harmful toxins therefore deciding when the fan should speed up and by extension it can always stay at the right setting remotely. All of the following list of technologies and related features are all to be considered when building designing producing and buying air filter and purification systems. Air is something none of us, and overall, no living creature can survive without and so constant examination, contemplation, and meditation should be taken into the processes in which we make our air clean and safe to breath. This all starts from the traditional methods of preserving the environment and being stewards of our natural air purification systems as well as using human ingenuity to provide solutions to problems using energy efficient technologies (Lafond, 2020).

CHAPTER 5

Environmental Remediation Technologies

CONTENTS

5.1. INTRODUCTION

Environmental remediation is an area that handles the extraction and removal, and disposal of contaminants that are present in the environment for example the soil, groundwater, and surface water. The term remediation in itself is defined as the action of reversing or stopping damages caused to the environment. This is the removal of source or breaking the pathways reduces exposure. The remediation process is highly dependent on many different conditions and basic standard requirements and it could also involve the assessment of the incremental health risk in situations where legislative demarcations have been set. The most efficient set of preliminarily remediation achievements in the United States of America comes from the Environmental Protection Agency (EPA). The European Union have also started embracing this technology by setting the Europe-wide standards, bearing in mind that majority of the industrially developed countries in Europe have developed their own standards with regard to environmental remediation technologies. Looking into Canada, majority of the developed standards for remediation in different provinces are individual for these provinces. The Canadian council of ministries in the environment which develops and sets guidance at the highest level in the embodiment of Canadian Environmental Quality guidelines and also the Canada-wide standards for petroleum hydrocarbons in soil. When you look at site investigation and assessment, immediately a site has been reported of being terminated, the automatic need to evaluate the grounds. There are usually assessments done and they begin with the preparation of Phase 1: Environmental Site Assessment, followed by Phase 2: Detailed Site Characterization, Phase 3: Consideration of Alternatives, and Phase 4: Remediation (Sara, 2003).

For example, historically, the development of many nuclear activities started without proper thought and consideration of their impact on the environment and surrounding communities. As a result of this, the birth of radiologically contaminated sites began. These sites have also been created by radiological accidents and also nuclear accidents as well as by non-nuclear industries by which human activity has greatly increased the exposure potential from naturally occurring radioactive materials to the unaltered state. Areas that have been contaminated could ultimately result to great health consequences for local people in communities, which requires appropriate action to be taken. Other instances resulting in contaminated soil/water include illegal dumping, lack of awareness of the impacts of toxic chemicals, poor containment in landfills and surface impoundments.

Several remediation activities were run in situations and circumstances where legal regulations and standards had not started to exist and in the event they were present, they were not taken seriously, or they were not comprehensive enough.

The introduction of the remediation process for contaminated land areas or contaminated environmental media can be done in two ways:

- The application of action to the contaminating agent itself. This in most cases can result to isolation, immobilization, and or the removal of the particular sources of contamination, for example by contaminating the area, the surfaces and the environmental media itself.
- Evaluation of risk related contamination exposure with regard to the population as well as also coming up with ways that can break the links between the contamination source to the people themselves. In most cases this particular approach leads to evacuation, the area is isolated, and or changing of the land use, as well as the local population's living habits.

These two ways are regarded complementary. In the process of deciding on the actual remediation work, there are several different factors that need to be taken into account. Every site has its own characteristics therefore there is no quick fix to these situations. One of the most important things is always to understand the action of remediation and also the need to be justified and optimized in order for these actions to be adopted to do better than harm. For instance, in the event of increased radiation levels, this does not bring us to conclude that this increase is harmful, due to existing environments having unnatural innate radiation levels got are high. The solution for this particular problem would be excavating or isolating the area in the absence of firm scientific grounds because it can subsequently cause a lot of distress to people that are concerned. Restoring a site that is contaminated to its initial state prior to contamination is sometimes not required or even achievable. Environmental remediation is concerned with reducing contamination exposure in order to keep the population safe. Remediated sites are always used for various activities of business such as industrial opportunities, and in some cases housing after remediation (Pichtel, 2019).

In the first phase of site assessment, the history of prior use of the site in question as well as the materials used and produced, help dictate the trajectory of the assessment strategy as well as the type of samples to be taken as well as the chemical analysis that will be conducted. Usually, the sites nearby that are owned by the same company and or in most situations reclaimed by the same companies as the contaminated grounds, are likely always contaminated in the same level as the original contaminated sites. It is also very important to take into consideration off site contamination of the sites nearby that is mostly through gradual and continuous emissions into the soil, air, and the ground water. The topsoil, surface, and ground water and the ceiling dust of the nearby properties and areas to the contaminated sites, should be taken through the process of testing which should be done pre and post any action of remediation (Pichtel, 2019). This is controversial requirement because it evokes different reactions by those responsible of these sites in that;

Nobody Would Agree to Pay the Site Clean-Up Charges: In the event there is the nearby properties are found to be contaminated, this will be registered, and it will appear on the owners' title of the property, which affects the value of the property.

Nobody is Willing to Take up the Charges of Assessment: It is common for many companies and businesses that deal with the voluntary assessing of the contaminated areas are always protected from the afterward paperwork sent to the functioning environmental agencies which is to be considered public as a result of the Freedom of Information Acts, which however a "Freedom of Information" inquiry usually attracts the presentation of related documents which are not safe guarder or will bring forth references to the paperwork done. Remediation should not be taken or confused with an emergency response after the occurrence of an accident because remediation usually does not require urgent actions. For this fact, there is need to conduct a thorough evaluation of the situation and the formulation of desired goals is not only possible, but it is a prerequisite. The proper planning is critically essential for any remediation work to be undertaken in order to reach the justified and optimized end results of the site in question. The aspects to take into account in thc dccision-making process differ in various angels of the process (Speight, 2019). These are economic, technological, and the social considerations, such as:

- The contamination risk posed to the people in the land as a result of the land use which is derived from the exposure assessment of the people themselves and the contaminated material in the site.
- The occupational exposure as a result of the remediation activities- for the workers that will be exposed to the contaminated material

during the remediation work.

- The net benefits of the remediation activities to the population affected.
- The waste generation from the remediation activities as the works are generally waste generating therefore warranting the consideration of the amounts and the properties of generated wastes in the decision-making process.
- The ethical issues involved as remediation may affect individuals lives, livelihood, and their living environment.
- The financial costs of remediation as it is a generally expensive process and there needs to be a well thought out mechanism for the payments.
- Other non-contaminated risks incurred in the remediation activities taken in the sites as some of the sites may not necessarily be contaminated by radio nuclides but also by other non-radioactive substances for example heavy metals and other hazardous organic compounds.

In the United States of America, there has been a mechanism that has been put in place for the taxing of the polluting industries which is used in the formation of a superfund to test and treat the abandoned areas. These funds are also used in the litigation to force companies and businesses to have their contaminated sites tested and remediation activities done. In different nations, there are similar regulations and often sites are rezoned to "higher" uses such as high density housing, which provides land a higher value from the original value in order for the developers to get an incentive from the funds after the costs of remediation have been deducted, which is used as money to purchase the land, clean it up, redevelop the land and eventually sell it and in most cases, in form of apartments (IAEA, 2012).

Remediation technologies are many and they vary depending on the circumstances, but they can also be classified as in-situ and ex-situ categories. The ex-situ type involves the excavation of the contaminated soils and an eventual treatment over the surface and also the removal of the groundwater that is contaminated and the treatment of the surface. The in-situ method is aimed at the treatment of the contamination taking away the process of removing the soil or the contaminated groundwater. There are also various methods developed to fit the remediation of soils that have been contaminated by oil. The process of remediation consists of the excavation of the soil and then disposing it in landfills and ground water is remediated through a technology called "pump and treat." In-situ technologies include solidification and stabilization, soil vapor extraction, chemical oxidation, Steam enhanced extraction, bioremediation-phytoremediation, Permeable reactive barriers and monitored natural attenuation though they are not limited to only these mentioned technologies. The technical processes that have been employed in the past with regard to remediation can be classified into three groups; these are physical, chemical, and biological technologies (Hussain, 2020). The various techniques that are used frequently are:

Containment;

- Pump and treat;
- Extraction;
- Stabilization also called solidification;
- Soil washing.;
- Nano remediation;
- Air stripping;
- Precipitation;
- Thermal desorption; and
- Bioremediation.

These are the main technologies that are applied as remediation technologies that this paper will look into and describe the way they work.

5.2. CONTAINMENT

Containment is a common technique that is adapted for the containment of contaminated ground water or the soil within a given site or a site that is in the same area as a contaminated site. The purpose of this is to limit the contact with the population in the area, the eco-system as well as reducing the chances of more contamination from happening. The main purpose of containment is reached when there is a successful setting up of low-permeability or impermeable cutoff walls. It is very common for the wall to be constructed with the soil, bentonite, and a water mixture. These kinds of walls house a blocker with low permeability can hold strong resistance at a very low cost. There are some welcome positions such as cement underwater may be needed as an addition for structural strength or in the event that chemicals are incompatible with the site contaminants (Hussain, 2020).

Wall construction is often done in conjunction with the simultaneous combination of excavation of the trenches, backfill mixing, support slurry fills and backfill emplacement. Depending on the sites condition and contaminants, this determines the trenches structure which can either be removed to a given level below the water table which captures the chemical "floatres" or it is driven deeper to a lower confining layer. One horizontal plane, the slurry walls are in a position to even be built to cover the entire premises of the contaminated area or they can be constructed in sections around the contaminated site. Soil-bentonite walls are very popular and also very cost efficient. Attapulgite is at times also incorporated in situations of bentonite being incompatible with the waste. In most recent developments the processes is the application of fly ash as high Carbon additive which not only lowers the permeability of the soil bentonite, but it also increases the absorption capacity with respect two organic chemical transport. Modern technology has stretched to accommodate the use of micro fine cement, mineral wax, sodium silicates as well as colloidal silica gel. The soils or the slurry walls permeated with a combination of these additives bring results of immense permeability lowering (Hussain, 2020).

The application of composite slurry walls as an additional barrier within the trench times improves the impermeability and immunity as well as resistance to the chemicals. The geomembranes are generally most times made of plastic materials which are compressed of high-density polythene pile plank sections. The membranes are always easy to install and have a long life, and the resistance applies to animals and vegetation intrusion and decay. This is very effective in constructing slurry walls and later on installing the membrane in the middle of the wall. The slurry walls have great potential to provide an effective elongated, and cost-efficient low maintenance solution in the control of contaminants migration in soils and groundwater. These types of walls are ordinarily quick when it comes to construction and provide a fast response solution in regard to accurate groundwater contamination problems. That being said, slurry walls also do not eliminate the source of the given contamination neither do they reduce the level of toxicity, and additional techniques and technologies are needed in the elimination and reduction of the possible threat of contamination (Pichtel, 2019).

5.3. PUMP AND TREAT

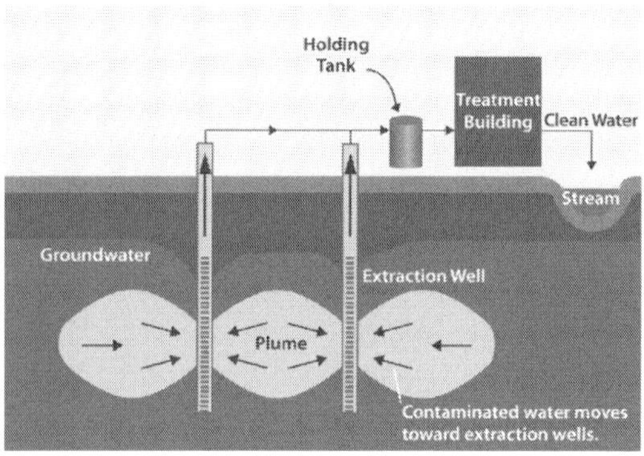

Figure 5.1: Schematic of the pump and treat method of groundwater remediation.

Source: https://geoforward.com/what-is-remediation-phase-2-environmental-site-assessment-phase-ii-esa/.

This is a key process for groundwater as well as soil remediation. It starts with groundwater being pumped to the surface and later on the pollutants are extracted in a variety of treatment processes that are inclusive of air stripping, activated carbon, ultraviolet (UV) or ozone treatment and precipitation. It involves the pumping out of the contaminated groundwater with the help of a vacuum pump which allows the extracted groundwater to be purified and gradually processed in a series of vessels containing materials that are designed to absorb the contaminants from the groundwater. In the case of petroleum contaminated sites, materials such as activated carbon is found in granular form. Chemical reagents are used to decrease the contamination levels of the groundwater (Figure 5.1).

Depending on the geology as well as the type of soil, pump, and treat maybe a very good recommendation to quickly reduce high contamination of given pollutants. At times it is very hard to access sufficiently low concentrations that's are satisfactory to the remediation standards because of the balance of absorption and disposition in the soil. Pump and treat is generally considered not to be the best form of remediation. It is considered pricy in the treatment of groundwater and also very slow in cleaning up a release with pump and treat. It is most suitable in the control off hydraulic gradient and keeps a release from spreading. There are more effective and applicable options in the same category as pump and treat such as soil vapor extraction, or dual phase extraction which is also called the multiphase extraction. Some other methods include the attempt to increase the dissolved oxygen that is contained in the groundwater to support microbial degradation of the compound through the direct injection of oxygen into the subsurface or the slurry which slowly releases of oxygen over time (Hussain, 2020).

5.4. SOIL VAPOR EXTRACTION

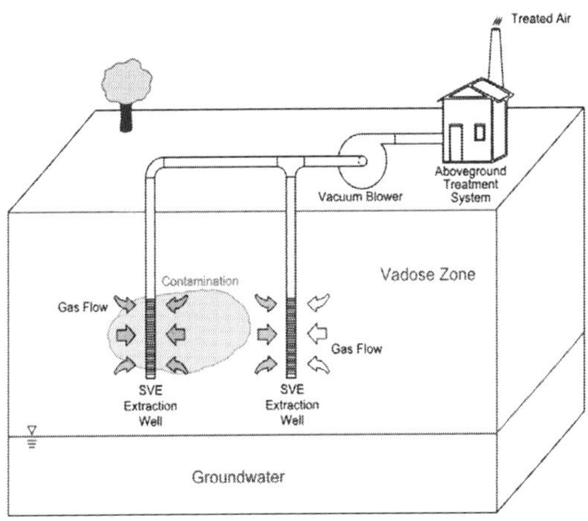

Figure 5.2: Image showing the soil vapor extraction and air sparing image.

Source: https://www.enviro.wiki/index.php?title=Soil_Vapor_Extraction_(SVE).

This is a successful remediation process that is used for soil that is contaminated. Multi-phase removal of the contaminated environmental features can be said to be successful with soil and groundwater. These two methods make use of different processes in treating the off gas volatile organic compounds which are produced as a result of a vacuum removal of air vapor from the subsurface and are inclusive of granular activated compounds, thermal, and or catalytic oxidation, and vapor condensation. In general, carbon is incorporated for concentration levels that are low in vapor steams, oxidation is essential for moderation concentration steams, and vapor condensation is required for high condensation vapor steams. Soil vapor extraction uses wells in the removal of volatile pollutants that are present in the soil. Vacuum blowers are usually built in as the primary force through the inducing of air flow through the soil matrix. These extraction wells are found within the polluted area. The wheels are generally made with a plastic pipe that is screened through the zone of contamination. The screened pipe is carefully inserted in a permeable packaging, the unscreened part is thereafter closed off with cement and this is to reduce the possibility of a short circuit airflow taking place direct to the surface. The volatile compounds are removed from the air which is all from the contaminated soil and these compounds are taken into the screened extraction well. The glasses are put together in a pipe or trenches with holes in it which are later moved above the ground into a separator that then on separates the vapor and the liquid (Hussain, 2020) (Figure 5.2).

The contaminated vapors are thereafter moved by the vacuum blower for the vapor treatment process. The vapors produced by this process are treated but carbon adsorption or thermal destruction. This is combined with other techniques such as biological degradation and UV oxidation applied to a limited extent. In some cases, the injection of air into wells are installed in-situ. These wills at times enhanced process efficiency by actively applying forced airflow. The system must always be formulated in a way that the air that is injected into the system prevents any escape of the volatile organic compounds into the atmosphere. The steam is often used in the extraction to remove the volatile hazardous pollutants present in the soil and the groundwater. It is usually injected into the ground and the soil temperature and also drives off volatile pollutants.

The steam injection also forms a displacement front through its condensation to displace groundwater. Steam can also enhance the stripping of volatile contaminants from the soil and can also be used to displace contaminated groundwater in the same situation.

The steam extraction process can be applied to organic waste, but it has not been applied in the removal of insoluble inorganic on metallic contaminants. The contaminated liquid's and steam condensate and are collected for further treatment (Hussain, 2020).

5.5. STABILIZATION/SOLIDIFICATION

The solidification or stabilization method utilizes chemical is reactive formulations that form stable solids which are non-hazardous or are less hazardous than the original materials. Solidification can be defined as the physical changes in the contaminated material with the addition of a certain binding agent. The different changes include the increase in compressive strength, A decrease in permeability as well as the condensing of hazardous materials. Stabilization on the other hand refers to the chemical changes occurring between stabilizing agents and the hazardous materials. These changes include less stable and less toxic materials containing hindered mobility. The common bonding agents that are used are inclusive of fly-ash, lime, Limestone, Portland cement, clay, and slag. Due to the various types of the hazardous materials, each agent is tested on the site before the commencing of a full-scale project (Hussain, 2020) (Figure 5.3).

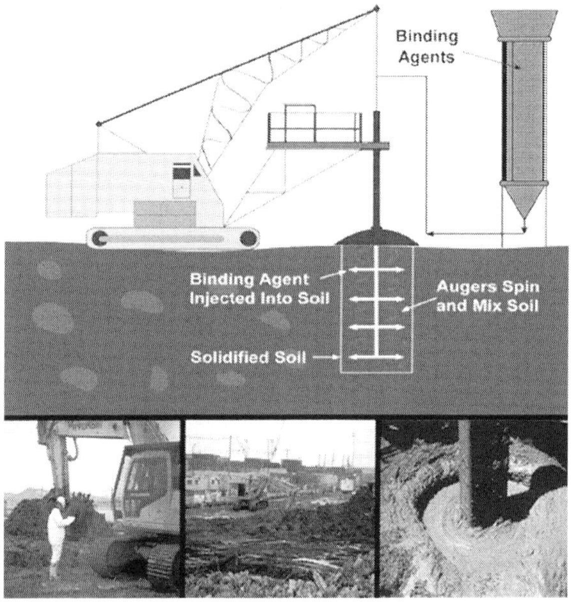

Figure 5.3: Stabilization/Solidification.

Source: https://www.researchgate.net/figure/Solidification-stabilization-of-contaminated-soils_fig4_260149770.

Majority of the building agents used are a blend of many single building agents all depending on the hazardous material that will be selected for the application. Portland cement husband used over the years to treat more contaminated materials in comparison with other materials under building agents because of its ability to bind free liquid's, a great reduction in permeability, and the reduction of toxicity with certain contaminants. Lime can be used in the adjustment of pH levels of the substance of drive off water by the application of high heat, a term referred to as hydration. Limestone can also be used in the adjustment of pH levels. Slag is usually applied for economical purposes due to its cost-efficiency. The prolonged use of the treated site, environment, and weather conditions could cause the materials used in the clean-up to stabilize the contaminates to erode which in turn limits the effect of the stabilization of the hazardous materials of contamination. Environmental factors like freezing-thawing and wetting drying are considered the focus of various studies that deal with the efficiency of stabilization as a remediation method. It was later found that freezing-thawing has the most adverse effects on the durability of the materials that have been treated. Dealing with radioactive contaminants, this remediation method is considered to be interrupted by an array of hazardous waste materials. Most stabilization/solidification processes very limited effects on organics and pesticides (IAEA, 2012).

That being said, the taking up of the solidification/Stabilization technologies of remediation have been considered relatively modest and a number of obstacles that have been identified, such as:

- There is the relatively low cost and widespread application of disposal to landfill;
- There is a lack of authoritative technical guidance on solidification or stabilization;
- There is uncertainty over the durability of the contaminant as well as the rate of contaminant release for the stabilized or solidified treated materials;
- The experience of past poor practices with the use of cement stabilization process that is employed in the disposal of waste in the 80's and the 90's; and
- The eventual liability that is linked with immobilized contaminants remaining on the sites in comparison to their removal or destruction.

5.6. SOIL FLUSHING

The soil flushing method is an in-situ remediation technology that is applied with the use of an aqueous solution to leach various contaminants form the contaminated soil into the solution used. The basic way it works is the application of water to the soil. The solutions used in the water are added to enhance the contaminant solubility and dissolve the contaminants into the flushing solution that is leached into the groundwater, which is extracted, and it receives treatment through this method (Figure 5.4).

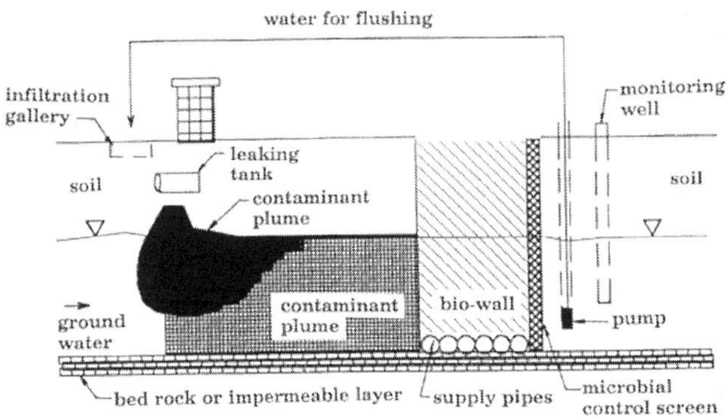

Figure 5.4: A soil flushing system.

Source: https://www.researchgate.net/figure/Soil-flushing-system_fig5_237228322.

This technology can be applied in the treatment of soil in vadose zones. In this case, the flushing solutions are sprayed onto the ground surfaces and they infiltrate throughout the contaminated area by the gravitational force all the way to the ground water. The leachate is thereafter removed by the groundwater into the extraction well. This remediation method can be used for many different types of organic and inorganic contaminants. Variables that affect the application of this technology are;

- The depth of the contaminated zones.
- The volume of concentration and contamination.
- The contaminant distribution coefficients between the soil particles as well as the flushing solutions.
- The presence of geological heterogeneities within the soil horizon.
- The interaction of the flushing agents/solutions with the contaminated soil.
- The suitability of the contaminated site for the installation of wells the delivery and recovery of the flushing solutions.
- The design factors include Sizing of the delivery and recovery systems which ensures complete recovery of the elutriate.

There are limitations and concerns that are attached to the soil flushing remediation method. There's a possibility that flushing additives could leave small residuals in the soil or the groundwater and they ought to be evaluated on a site-specific basis. Flushing could also spread the contaminants beyond the capture zone, or they could introduce surfactants onto the subsurface. This requires the technology to be used only where flushed contaminants as well as the flashing solutions can be contained and recaptured. The additives might also have to be recovered from the underlying aquifer and impossible occasions they should be recycled. The treatment of recovered flashing solutions from the residual sludge must be disposed of. The recovered groundwater might need prior treatment to meet the appropriate discharge standards. In the event volatile organic compounds are extracted; the air emissions should be treated. Soils with low permeability such as clay are often very difficult to treat using this method. The remediation time baby long and when the contaminated zone is very deep, the flashing solutions can incur a lot of charges which extends the remediation time and process (Speight, 2019).

However, the advantages of this technology include;

- It is an in-situ technology that causes less exposure of those involved in the clean-up process and the contaminated environment;
- It is relatively considered a simple and cost-efficient operation to implement;
- It can be applied in a wide variety of contaminants which fall under organic and inorganic contaminants;
- It can be applied in situations such as saturated and unsaturated contamination zones; and
- It can be pad with many other remediation technologies to achieve the cleanup goals.

5.7. NANOREMEDIATION

This is the primary function of nanoparticles in the purpose of environmental remediation. It has been investigated in the treatment of groundwater, soil, wastewater, or other contaminated environmental materials. This remediation method can be considered an emerging industry. By the year 2009, Nano remediation processes had been recorded in up to 44 Remediation cleanup sites all over the globe and especially in the United States of America. In Europe, this technology had been investigated by the EC funded Nanorem project. During Nano remediation, the nanoparticle agents have to be mixed with the contaminant under conditions that support detoxifying reactions. This activity usually facilitated by a pump and treat technology prior or an in-situ application. Some of these methods, in particular the application of Nano zero-valent iron for the cleanup of groundwater have been deployed at a cleanups site. This method is however still under research and development to achieve better results and to meet the remediation standards (Pichtel, 2019).

This technology has been widely used for the treatment of groundwater with the addition of extractive research in the treatment of wastewater. In groundwater remediation, the application of nanomaterials is a continuing upcoming approach that shows promise as a result of the immense accessibility and the success of the nanomaterials that are used together with the contaminants. Nanotechnology presents great potential do effective treatment of contaminants in situ, which avoids extraction or the requirement to pump the polluted water out of the ground. This process starts with the nanoparticles being injected and mixed with the contaminated water in the aquifer through an injection well. The nanoparticles thereafter transported by the groundwater flow into the source of contamination. Just as contact takes place, the nanoparticles can isolate or hideaway the contaminants which immobilizes them, and the degraded contaminants are considered to be less harmful compounds. Redox reactions are considered to transform the contaminants. It is considered reactive when the nanoparticles are the oxidant. The ability to inject the nanoparticles and to transport them to the source of contamination is crucial for the success of the treatment. The reactive nanoparticles are injected into a well where they are transported into the contaminated area. The drilling and packing over well can be very expensive. Direct push wells are considered to be cheaper done the drilled wells and are also often used as a delivery tool for remediation with Nano irons.

In surface water treatment the application of various nanomaterials provides a lot of promise for the successful treatment of surface water including purification, disinfection, and desalination. The contaminants that are targeted in surface water treatment include organic contaminants, pathogens, and heavy metals. Nanoparticles can also be very important in the detection and tracing contamination levels in contaminated sites which is key to effective remediation. Nano remediation materials are considered to be highly reactive due to the fact that they have higher surface area per unit mass, and also this reactivity nanomaterial can also react with target contaminants quicker than they would with larger particles. In many Nano remediation technology application, the use of Nano zero valent iron is done in most fields with the mixture of other metals enhancing dispersion. The nanoparticles are considered highly reactive and they can quickly clump up together or react with the soil particles as well as other materials in the environment which is considered to limit their disposal to the target contaminants. Some of the challenges that come with this technology is that when using the in-situ remediation the reactive products have to consider two reasons. First, the reactive compounds cannot times be more harmful or mobile that the parent compound. Second is that the product may affect the effectiveness or the productivity as well as have an effect on the cost of the remediation process. Another limitation is that identifying coating or the formulation that's increases the spread of the nanoparticles when they come into contact with the target contaminants while also limiting any possible toxicity of the bio-remediation agents such as wildlife or people (Pichtel, 2019).

5.8. AIR STRIPPING

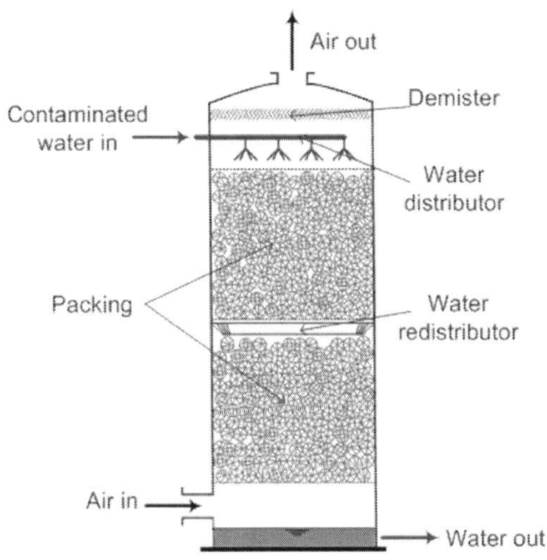

Figure 5.5: Air in a stripper.

Source: https://en.wikipedia.org/wiki/Air_stripping.

Air stripping can be defined as a transfer of volatile components of a liquid into air vapor. It is considered to be an environmental engineering technology that is used in remediation of groundwater and wastewater that contain volatile contaminated compounds. The volatile compounds have very high vapor pressure and they also have low aqueous solubility which is characterized by the compounds Henry's law coefficient. Volatility is equal to the tendency of a given compound to evaporate when it is under normal atmospheric conditions. The influence of Henry's law, just like volatility describes the ability of a compound to transform from a liquid to a gas at equilibrium. This law presents a ratio of the constant equilibrium present in liquid phase and also the gas phase. Pollutants that have a relatively high Henry's law coefficient can be economically stripped from water. These pollutants include benzene, xylene, toluene, and ethylbenzene that are present in gasoline. Due to the fact that Henry's law coefficient increases with temperature increase, the process of stripping is made easier at these warmer temperatures.

Devices that promote contact between air and water seem to strip volatile compounds, but air strippers are packed towers and tray towers use the operation with countercurrent flow of water and air. This countercurrent flow tends to remove particles from the water and deposits them in the air. This process is called volatilization or air stripping. The water is deposited into our system from the bottom. When the water reaches the bottom of the system it is usually considered treated although additional testing can be done to determine whether the standards of radiation have been met. Since there are many of contaminated compounds stripped from the water making the air contain hazardous pollutants the air exits the stripper and it may require emission control. Carbon adsorption is most commonly used and catalytic oxidation is another consider the option. There are two different types of strippers that are mainly used. One is the packed tower systems and the other sieve tray system. These different types of strippers are used depending on the types of contaminants as well as the level of contamination in the water source being extracted (Pichtel, 2019).

The packed tower systems, as depicted in figure 5.5, make use of a distributor that is placed at the top of the tower which ensures even distribution of the engineered plastic, ceramic, or metal packing which maximizes the air to water contact. The design criteria for a packed tower includes the air to water flow rate, the surface area which is provided by the packing, as well as the column height and diameter. The designers of air strippers aim to gain the largest possible air to water surface contact increasing the maximum efficiency and removal of the contaminants from the water. Packed strippers can be considered very efficient in the removal of low volatility organic compounds even better than tray strippers. In addition, packed tower strippers are considered to be very cost efficient when there is treatment of large quantities of water. These towers can be between 5 and 12 meters high and they are also a permanent installation. There is a possibility that some towers can be transported on moveable trailers and used in a different contaminated area well water treatment is required. Trade towers are also called sieve tray towers. These towers use the same process as we find impact towers, though instead of having packed materials being distributed evenly, the materials to be used as separated into different trays and have holes that allow the dripping of water through them. The use of an electric air compressor is positioned at the bottom of the system where the air coming from the fans goes through the holes and is exposed to the water. Unnatural draft can also be introduced as an air source which separates the contaminants from the water. This natural draft is used for the removal of a bit more volatile substances as compared to packed towers. Mechanical air compressors are also used in the removal of less volatile substances in water. With regard to the amount of time it takes for the filtration of water through air strippers, this time can vary considering the system that is used Anne also depending on the size of the tanks, as well as how quickly the water flow Is through these devices. The estimated amount of time that water takes to be filtered is around a handful of minutes. That being said, there are studies that have suggested that it can take longer depending on type and the concentration of the contaminating substances (Pichtel, 2019).

5.9. PRECIPITATION

Precipitation has become one of the most widely selected means of remediation in the removal of heavy metals from groundwater in a pump and treat operation. It can also be a primary method in the treating wastewater from industries that contain heavy metals. As a result of this methods success, this technology was approved in the remediation of groundwater containing heavy metals which include their radioactive isotopes. In ground water treatment this technology is used as a pretreatment for other treatment technologies to be applied such as chemical oxidation and air stripping where the presence of the contaminating metals would interfere with the other treatment processes.

The precipitation process normally adjusts the pH levels, it also adds can recalls that stimulate precipitation, the addition of coagulants and also the mixing of fluids in the devices used in precipitation. These devices are known as flocculators. This time refers to the quality coalescence of particles. The coagulants, the chemical precipitant as well as the flocculator, all put together in order to increase the particle size through aggregation. Commonly the use of metal precipitates from the solution as hydroxides, sulfides or carbonates. The process used is dictated by the solubility of the specific metal contaminants under required cleanup standards. In some cases, the progress design allows the generation of sludge is which can be sent to recyclers for the recovery of the metals (Hussain, 2020).

5.9.1. Coagulation and Flocculation

As previously mentioned, the Coagulants, the chemical precipitant as well as the flocculator increase the particle size through aggregation. It is normal for the precipitation process to produce very fine particles which are suspended and collected by electrostatic surface charges. These charges bring about clouds from counterions and form around the particles giving rise two repulsive forces which prevent aggregation and reduce the effectiveness of liquid to solid separation. The chemical coagulants are there for added in order to power the repulsive forces that formulate around the particles. After addition of the coagulants, low-sheer mixing in a flocculator promotes the contact between the particles and allows the particles to grow through the segmentation phenomenon known as the flocculant settling. In an anaerobic environment, the bacteria react with the soluble metals to form insoluble metal precipitants (Hussain, 2020).

There are limitations and concerns expressed by the use of this remediation process. Precipitation is considered a difficult process in the purification of contaminated water with multiple metals. Precipitant that have been obtained from the cleanup process have to be tested for leaching before land disposal due to the concern that after the disposal, the metals leach from the precipitants due to the different changing environmental conditions such as moisture and pH fluctuation. If the precipitation process is geared after removal of radio nuclides, the amounts of these radio nuclides that can be extracted all depend on the precipitant, the pH of the solution, and the concentration levels of the radio nuclides present in the contaminated water. With regard to the addition of chemicals, this must be done carefully, and it must be controlled so that the unacceptable concentrations in treatment well not occur. The discharged water after treatment usually requires pH adjustment (Pichtel, 2019).

5.10. THERMAL DESORPTION

This is an environmental remediation technology that takes advantage and uses heat to increase volatility of the contaminants paving way for them to be removed or separated from the solid matrix. This technology uses heat contained in soil or sludge in-situ or ex-situ Which Is used for the removal of contaminants from soil. Volatile and semi volatile organic compounds I removed from the soil in thermal desorbers at a temperature of 100 to 300°c in those desorbers that are low temperature. For the desorbers that are high temperature desorbers, the temperature rages from 300 to 500 °c. The vapors formed enter a gas treatment system where they are treated. The gas treating devices are at times taken to be considered condensers which filter for organic compounds that are to be treated and disposed of. These units can step in as catalytic oxidizes that destroy the organic contaminants all the way. The basic contaminants of this remediation system include;

- Air emissions control;
- Thermal processor on discharge; and
- Pretreatment and feed.

The critical factors that determine the effectiveness of the application of this technology include;

- The type of Soil and the amount of moisture content;
- The temperature;
- The residence time;
- The mixing; and
- The sweep gas flow rate present in the thermal processor.

This technology is considered very successful in the remediation of soils that have oil contamination, coal tar waste, wood-treating wastes, creosotes, fuels, PCBs, mixed wastes, synthetic rubber processing waste, chlorinated solvents, hydrocarbons, pesticides, and plant waste (Figure 5.6).

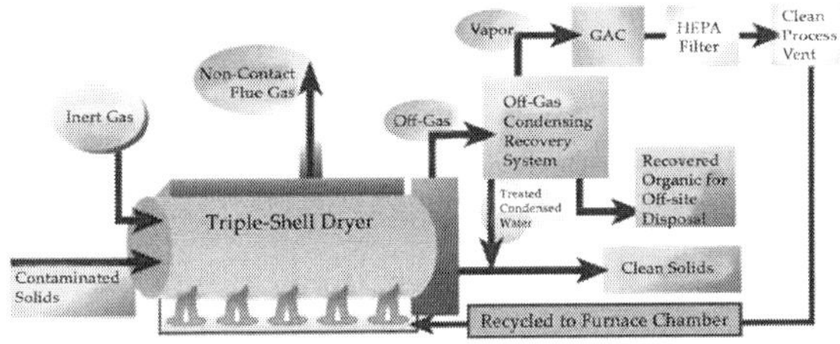

Figure 5.6: Thermal desorption.

Source: https://www.geoengineer.org/education/web-class-projects/cee-549-geoenvironmental-engineering-winter-2013/assignments/thermal-desorption.

Based on the operation of temperature in these desorbers, this technology can be categorized into two groups. We have the high temperature thermal desorption (HTTD) and the low temperature thermal desorption (LTTD). It is very critical for one to note that this technology does not destroy organics. In HTTD, the wastes are heated to Temperatures of up to 560 degrees Celsius. HTTD Ah frequent Lee used in combination with other processes such as solidification and stabilization, dichlorination which all depend on the site. In LTTD, the wastes are also hated but at a lower temperature of 90 to 320 degrees Celsius and is most often used in the remediation of soil that have fuel contamination. Unless the temperature range of the LTTD is increased to the higher end, the organic components present in the soil are not damaged at all and this allows the treated soil to retain its ability to sustain future biological life or activities thereafter (IAEA, 2012).

In the off-gas treatment, there are only three basic options available. The volatilized contaminants into the off gas can be discharged into the atmosphere or they can be collected, and we can also be destroyed. In a number of cases, both the collection and destruction system were employed. In addition to that, managing the volatilized components the solid particles that leave the desorbers, must be removed from the off-gas. In the use of a collection system, the off-gas is required to be cooled to condense the volatilized components into liquids. The off gas will remove most of the desorbers in the HTTD temperature range. This cooling father initiates the condensing of the volatilized water inorganic contaminants. Due to this reason, after the condensation process there is a required further treatment of the off-gas. The cold of gas may be treated by thermal oxidation. Thermal oxidation is usually attained by using catalytic oxidizers to their combustion heat source of the desorber. The volume of the gas that needs treatment for indirect fired desorbers is only a fraction of that which is required for the direct fired desorbers. (National Service Center for Environmental Publications, 2020) Available at: shorturl.at/cdmpF This needs small amounts of air pollution control in the emissions released. Some thermal desorption systems recycle the carrier gas which further reduces the volume of gaseous emission (IAEA, 2012). Limitations that come with the application of this technology are as follows;

Dust and organic matter in the soil are increased as a result of treating the off-gas.

- Leaching Mercury from stockpiled soil into the water is an issue of concern especially for the population that indulges and depends on fishing.
- Heavy metals present in the feed have a possibility to produce a treated solid residue that needs stabilization.
- Play under the soils present with high humic content requires longer reaction time.
- Dewatering might be very necessary in the achievement of acceptable soil moisture content level.
- The treated soil could have no support for microbiological activities which breaks down the contaminants. In the event that the soil is returned to its original site this might be a matter of concern to the population present.

5.11. BIOREMEDIATION

Bioremediation is a process that is used in the treatment of contaminated water, soil, and subsurface material through altering the environmental conditions which are suitable from growth of microorganisms and bring down the existing target contaminants. Bioremediation is considered less expensive in some situations and also most suitable in comparison with other remediation technologies. The biological treatment applied in similar approaches would be used in the treatment of wastewater, industrial waste, and solid waste. Many of the bioremediation activities are inclusive of oxidation reduction where an electron accepter he's introduced with the aim of increasing oxidation over reduced pollutant such as a hydrocarbon or electron donor is introduced to reduce oxidized pollutants. In this above approach, the addition of nutrients, minerals, vitamins, and pH manipulation all optimize the conditions for the microorganisms to prosper. In some circumstances, special microbial cultures Are introduced to further enhance biodegradation.

Most of the bioremediation processes involved redox reactions that require chemical species to donate an electron to a different species which accepts the electron. During this process, the electron donor is it regarded as the oxidized agent while the electron acceptor is reduced. Some common electron acceptors in this particular technology include oxygen, manganese III and IV, carbon dioxide, nitrate, and sulfate. Electron donors in this process include sugar, alcohol, fat, and natural organic materials. In aerobic bioremediation, it is very common from oxidative bioremediation processes in places where oxygen is given as the electronic acceptor for the oxidation of petroleum. Oxygen is generally preferred to be used as an acceptor because it yield a little bit more energy and also oxygen needs a number of enzyme systems to start the degradation process. Common approaches which provide oxygen on top of the water table include bioventing, composting, and land farming. On the process of land farming, contaminated soils or sludge are incorporated and added into the soil surface in a tilted manner using conventional agricultural equipment in the mixing process. Composting quickens the pollutant biodegradation by mixing the waste that requires treatment with the bulking agent which forms into piles and which mixed increase oxygen transfer. In anaerobic conditions, the support of biological activity by which no oxygen is present is necessary in order for the microorganism's breakdown chemical compounds taking place in the soil releases energy which the microorganisms need. At times during the aerobic and anaerobic processes of breaking down contaminants, intermediate products which are equal, less or more toxic then the original contaminant are created (IAEA, 2012).

The biodegradation of microorganisms is helpful in many types of organic waste and is very cost effective due to the fact that it is a natural process. Different techniques can be employed and conducted on site completely eliminating the need to transport hazardous material from one point to another. The extent of biodegradation solely depends on the toxicity of the contaminants and the initial concentration level of the same contaminant, the properties found in the contaminated soil, the particular treatment system itself, and the biodegradability of the contaminants present in the soil. In the cleanup of heavy metals, the heavy metals such as lead, chromium, cadmium, and uranium can all not be biodegraded. That being said, bioremediation can potentially be applied in the reduction of these materials in the subsurface, reducing the great potential for human an environmental exposure also posed by these materials. The mobility of certain materials all depends on the oxidation state of the material in question. If there is a possibility of bio stimulation, the addition of nutrients that are limited to making the environment more favorable for biodegradation can be taken into consideration and implemented to make the system more effective. Many biological processes are significantly sensitive to the pH and the function most effectively in the presence of nuclear conditions. Low pH levels can definitely interfere with pH homeostasis, which is also the increase in solubility of the toxic materials most preferably the metals (Speight, 2019).

Biodegradation can be applied to mineralize organic pollutants completely, to change the pollutants up to a certain degree, and or change their mobility. Heavy metals and radionuclides stand as elements that are not biodegradable, though they can be bio altered into less mobile forms. In a few situations, microbes cannot mineralize all the way to pollutants that leave a possibility for a more toxic compound to form. Biodegradation only occurs in the presence of microbial population with the metabolic capacity to completely degrade pollution, the specific environment with right growing conditions for the microbes at hand, as well as the correct amount of nutrients and contaminants. The biological processes that are required by the microbes in question are extremely specific, meaning that many environmental factors have to be taken into careful account (Speight, 2019).

5.12. COMMUNITY CONSULTATION AND INVOLVEMENT

In the preparation stages of a remediation activity, it is a requirement that there is extensive community involvement and consultation. The proponent Is required to provide adequate information to the community and also get information from the same community. The proponent also needs to get information on the sensitive uses and matters of the community such as their livelihood, schools, playgrounds, childcare, as well as hospitals and look into the community concerns, interests, and questions. The consultation process should be open and undertaken on a group basis in order for both groups with their member representation from the community as well as the proponent site can have an amicable discussion on the issue at hand taking into consideration both sides interests. An independent mediator who comes in as a chairperson should preside over these consultation meetings. It is crucial that the chairperson is approved and accepted by both the proponent and the community. It is required for minutes of every meeting to be taken down including the questions asked by the community members and the answers given, to be available on the internet as well as have it kept at a local library or a Community Center where everyone can access.

5.13. PROGRESSIVE HEALTH HAZARDS AND RISKS

This health risk is considered to be the vulnerability factor due to the exposure and the commencing of the remediation project in the community but could harm both members of the community and the workers in the remediation site themselves. The use of progressive health risks is based on the carcinogenic, mutagenic, teratogenic effects among others and are usually involving the value judgments concerning the estimated rate of growth in number of cancer infections that are acceptable. In different jurisdictions the incremental health risk rate differs from one to another. In some jurisdictions, it can be 1 in every 1,000,000 people, but in another jurisdiction, the acceptable projected incremental health risk rate can be as low as 1 in every 1,000,000. The incremental health risk present from one remediation activity is not of any comfort if the area that is contaminated has already be considered to have a high help risk from previous operations in the same area or if have a project exist at the same time as the beginning of another project. We can use the analogy used by different remediators which allows us to compare the risk of the remediation treatment process to the population in these areas and the risk of loss of one's life in form of a car accident, smoking of cigarettes, food poisoning, or a heart attack.

5.14. EMISSION STANDARDS

There are standards that are set to regulate the levels of noise, dust, odor, emissions released in the air, emissions released into the ground water, the discharge Released into sewers and waterways all containing chemicals. These are mirrored with the original levels in the area and the standards used in other existing or completed remediation processes that took place in a given are. In the event an area is considered an industrial zone and is permitted emission release, this does not apply to nearby residential areas that desire the same, unless they meet the required standards. The monitoring of these standards individually is done and taken very seriously by the local community as well as their designated authorities. There is a need for the partnership with law enforcement in the event these standards are breeched, and this hinders the repetition or the possibility of more breaches happening in the future. In the transportation section, in the event there is a need to transport contaminated material from one site to another, the proper protective measures should be undertaken. This is inclusive of the site staff transporting the materials, the community involved, and any other persons involved with the sites.

When it comes to the funding of these remediation projects, there is immense hesitation by the local community as well as the local government, due to the sole repercussions and effects that may emanate from this same process. They minimize their involvement in the event there are health issues that could develop in a span of days all the way to a span of years. This also detaches them from the effects such as dust, emissions, noise as well as the interruption of normal life in schools, childcare services, hospitals, playgrounds, and other public facilities (Speight, 2019).

In conclusion, there are very many different and available remediation technologies that can be applied in the treatment of the polluted soils, wastewater, and ground water as well as air. There are however a variety of limitations to the applicability of a particular method to a given site based on the various conditions on both sides. There is also the limitation of cost featured in some of the methods and technologies applied though there are those methods that have been proven to be cost-efficient and give very commendable results. With the combination of recent technologies, there are better safer ways of carrying out the remediation processes. This is ultimately for the better functioning and the effects that come from them to be reduced in number and potency. There is a need to educate the general public on these methods and technologies. In the event there is a remediation activity in the community, they need to be aware of their right to information on this activity and the ways by which they can keep safe. This comes with the safety of their children and families as they could be required to reduce contact with various materials that at times are a part of their daily lives.

There is also a greater need to implement laws that prevent businesses and companies of an industrial nature from releasing contaminants and substances into the environment, whether solids or liquids or in gaseous form. These habits build up and they lead to the contamination of the soils, water, and the air. With the increase of urbanization and industrialization where there is manufacturing of majority of the items that are used in man's day to day activities, the pollution levels go higher. There need to be serious regulations set in motion that reduce the production levels per day and if not, these companies incorporate environmentally friendly ways or clean ways for mass production of that which they produce. The increase in industrial emissions in all forms increases the climate change that is caused by environmental degradation. Warmer winters, extremely dry seasons, delayed rainy seasons, cyclones, earthquakes, locust invasions just to mention a few of the environmental disasters that are caused by the environmental degradation and climate change (Speight, 2019).

The implementation of these laws does not only help stop more severe damage to the environment, this helps with the creation and the setting in motion measures that will restore the environment. The planting of trees is detrimental to the purification of air and attracting rainfall in the areas that have less rainfall than others. Restoring the environment saves us as humanity a lot of damages in terms of property, resources, economically, agricultural loss, and the loss of human life.

The current times are very prone to cancer infections at a very high rate due to lifestyle issues. There is a given percentage of these cases due to the exposure of pollution especially in big industrial cities like Beijing and Baoding in China and Gaya in India, where the pollution levels bring about the air and water contamination health risks and health conditions. These health conditions are very difficult, and they can easily take over a population due to their costs in treatment and management. This is because most of the health conditions are still being studied by doctors as they are new, and they have no cures or treatments. This greatly affects the productivity of the population and the nation at large.

These are conditions and situations that can be avoided by law. Relevant legal authorities can start on a local level as we work on going higher into governmental laws and global laws by developed, neutral bodies started by environmental agencies set at achieving the ultimate goal of environmental degradation as well as avoiding remediation in general on a larger scale (Speight, 2019).

CHAPTER 6

Soil Conservation Technologies

CONTENTS

6.1. INTRODUCTION

Soil conservation defines a number of techniques that are used to preserve soil. It can also be referred to as practices that are employed in the protection of soil from degradation. Soil conservation involves the treatment of soil as an ecosystem that is alive. This means tending to soil in order to maintain its value or resources in a way that is sustainable. The process of soil conservation can be compared to maintaining vehicles, in practice vehicles require regular servicing which includes changing oil and filter, checking hoses, and spark plugs on a regular; all the above is done to prevent major repairs.

As mentioned above it is necessary to treat soil as a live ecosystem, this is done by recognition of organisms that inhabit the soil and their different roles they play in enhancing soil health and fertility. Soil contains biogeochemical reactions and interactions between several organisms that are alive like soil microorganisms including soil bacteria like actinomycetes, cyanobacteria, soil fungi, viruses, slime molds, lichen, and various soil protozoa. There also arthropods present, mites, springtails, root aphids, millipedes, earthworms, nematodes, snails, and slugs, pot worms, flatworms among others. Some of these organisms are plant pests that lead to infections of plants or crops but a number of the organisms are of benefit to the soil conservation and formation process. This makes the soil a dynamic ecosystem and a site for interactions of different micro, meso, macro, and megafauna that are dependent on each other for food and energy (FAO, 2000).

Some of the organisms are responsible for organic matter breakdown, nutrient release, and the creation of spaces to allow water and air circulation. Some of the organisms responsible for soil health and fertility require dead organic matter for food and energy. This means these resources need to be replenished on a regular and continual basis. Organic matter also provides good soil structure and improves the water holding capacity of soil. It also helps to promote water infiltration and protects soil from being eroded and compaction.

Applying tillage on agricultural land causes disturbance to soil flora and fauna and leads to the damage of vesicular Arbuscular Mycorrhizal fungi (VAM) which take a long time to establish their symbiotic relationship with the growth of crops. Leaving land fallow is a bad agronomic practice since it helps in soil erosion and impacts soil moisture and also soil temperature. Best practices involve leaving soil undisturbed as much as is practically possible to prevent the disruption of microbes that are beneficial to the soil and prevent damage to the soil ecosystem hence securing soil health in a manner that is sustainable. The practice of no tillage agriculture and covering soil with cover crops that are available during a period is important to protect the soil from erosion, loss of moisture and regulation soil temperature in order to maintain the structure of soil (FAO, 2000).

The first principle of soil conservation is to preserve soil life and organic matter. Other principles of soil conservation include managing surface runoff, protection of bare exposed spoil surfaces and sites that are highly susceptible and it also protects downstream watercourses from sedimentation and pollution.

Similarly, to any terrestrial and aquatic ecosystem, fauna present in soil on a higher trophic level are dependent on the one lower. Fauna in soil through complex activity that is of physical nature; supported by geological resources that are helpful in soil an example like minerals in soil. It creates hydration and creates aeration channels in layers of soil; creating air spaces for different organisms to foster the completion of their cycle in full and in part. Soils helps the root of plants to grow and access minerals as well as reaching the water table easily, it also assists in biodegradation, bioconversion, and transformation processes in soil. Any alterations made in the surface of soil, soil top layers and soil profile with the use of agricultural methods that are unscientific can lead to the breakdown of food chains deemed as complex and existing food wens in soil. It makes soil less productive, causes changes in the PH of soil, it can also lead to the generation of mutant soil fauna species and finally the destruction of cycles that are biogeochemical in nature and operate in soil. Soil conservation processes are therefore important to employ soil conservation practices in order to protect global resources obtained from soil. Unless comprehensive steps are taken to protect soil; then agriculture and forestry could be affected globally and face challenges that are substantial (Hellin, 2016).

The well-being of people can be dependent on soil health since it support the growth of food and therefore it contributes the economies of countries. Soil also acts as a filter cleaning water and air. It has functions such as exchanging gases with the atmosphere and therefore makes a contribution to climate globally. Soil also receives waste that is organic and goes on to recycle their nutrients back to plants. Soil also breaks down and holds waste that is toxic. Since soil plays a huge role in the health of the world, it should be conserved in a manner that is sustainable (FAO, 2000) (Figure 6.1).

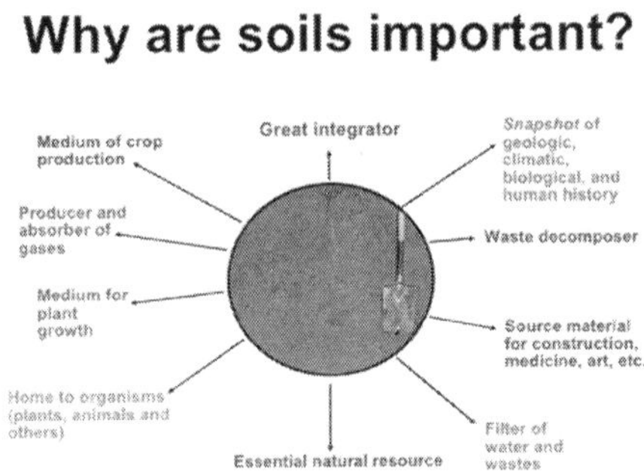

Figure 6.1: The importance of soil to the environment and ecosystem.

Source: https://empoweryourknowledgeandhappytrivia.wordpress.com/2015/01/29/soil-layers/.

There are different causes of soil loss and soil fertility they include over cultivation, erosion, salinization, and contamination by chemicals. Subsistence farming which is unsustainable and traditional farming methods such as slash and burn clearing methods used in some developed countries lead to the nutrient loss, deforestation, massive scale erosion and sometimes it eventually leads to desertification (Denboba, 2005).

Soil erosion leads to the loss of top soil that is necessary for organic matter, nutrients, microorganisms which are all necessary for plant growth. Soil conservation measures protects all these characteristics of soil necessary for plant growth. The soil that is eroded ends up in aquatic sources depositing pesticides, fertilizers used for agriculture. Taking necessary steps to conserve soil is considered environmentally friendly and can foster the achievement of sustainable development goals.

Soil conservation is a process that is active and ongoing and those practicing it must maintain their commitment in doing so. The first step in the soil conservation process is to obtain knowledge on the land resource. This means knowing and understanding where soil is considered to be most permeable and has the most susceptibility to groundwater contamination as a result of excessive use of pesticides; it also entails understanding where land is most susceptible to soil erosion combining the nature of the slope and the soil texture. Without understanding this, it will be impossible to prepare and plan for an appropriate strategy for the conservation of soil. Other steps follow and they include identifying and predicting problem areas, selecting, and implementing soil conservation techniques and maintain control structures. The final steps is the continuous monitoring of the effectiveness of the soil conservation plan and making changes in the event they are necessary. Several measures, techniques, and technologies that have been developed to help conserve soil by the use of good agricultural practices. Techniques for soil conservation include the rotation of crops, cover crops, conservation tillage, planting wind breaks. All the above techniques influence both soil fertility and erosion (Hellin, 2016).

6.2. RATIONALE BEHIND SOIL CONSERVATION/APPLICATION OF CONSERVATION TECHNOLOGIES

Soil conservation practices are used during operations of harvesting and replanting. Soil is a resource that is extremely valuable at the same non-renewable. Most of the time it has often been neglected and not given the attention it deserves. There are several factors that are anthropogenic such as environmental pollution, rapid industrialization and agricultural practices without any substantial environmental planning and the growth of the global populations that affect soil quality. Other reasons and factors include the destruction of natural virgin forests and jeopardizing ecosystems for developing infrastructure without any carrying out an impact assessment, political instability, socio-economic crisis, infringement in areas that are forested, excessive use of agricultural chemicals, land use in a manner that is not sustainable, overgrazing, use of agronomic practices that are not environmentally friendly, slash, and burn agricultural methods and wild fires just to mention a few (Hellin, 2016).

All the above factors are responsible for affecting the health of soil, soil fauna, physicochemical characteristics of soil such as texture, soil profile, soil pH, permeability, fertility, mineralization, porosity, deposition of soil organic matter, top soil loss as a result of extensive erosion, low productivity and a significant increase in soil borne illnesses all around the world. They are several reasons for practicing soil conservation. They include.

- Maintenance of organic soil matter and the biological life present in soil-these two soil components account for a large percentage of soil productivity by improving soil fertility and soil health.
- To ensure constant supply of food and at affordable and reasonable prices-overtime soil conservation has proven to increase the quality of soil and also the quantity of agricultural produce since

it maintains and keeps top soil in its place hence preserving the long term productivity of the soil.

- It increases global food productivity and improves global food security that can also be provided to other countries experiencing food shortages.
- It leads to economic gain as farmers can save money-the effects and impact of soil erosion can save farmers additional costs of up to 90 million dollars annually and much more, especially in relation to inputs used to increase soil fertility and reduced productivity of crops.
- It also saves citizens money which can be estimated up to 9 million years annually according to recently published research.
- It also improves the quality of water-all forms of aquatic life need clean water for their survival and this can be affected by agricultural activities especially agricultural inputs such as fertilizers and pesticides used in excess by farmers to improve soil fertility; it is also as results of soil erosion which leads to sedimentation and contamination of water resources which affect the aquatic ecosystem by challenging their survival and also their biodiversity.
- Improvement of wildlife habitat-practices to conserve soil such as the provision of buffer strips and wind breaks; replacement of organic matter increase and enhance environmental quality which supports wildlife.
- For aesthetic reasons-conserving soil on different landscapes makes land more attractive picturesque scenery.
- It also helps to create an environment that is free from pollution and thus making these areas safe to live or support life.
- For the purpose of sustainability, for future generations so that they have enough soil to support life for all biological species.

6.3. PRINCIPLES OF SOIL CONSERVATION

Erosion contributes to loss of soil health and fertility. This forms of erosion include water, wind air and water. The art of soil conservation is based on several principles which include:

- Protection of soil from the impact of rain drops;
- To slow down water movement and concentration down the slope into narrow paths;
- To slow water movement when it is moving along a path;
- To encourage water filtration in to the soil;
- To increase the size of soil particles;
- Reduction of wind velocity near the ground by growing vegetation cover, ridging the land among others; and
- To grow the strips of stubble or other vegetation cover which might catch and hold moving water from carrying soil.

Keeping the above principles in mind, environmentalists, and ecologists have devised different technologies which can preserve loss of soil during erosion.

6.4. TECHNOLOGIES OF SOIL CONSERVATION

6.4.1. Evolving Conservation Technologies

Agricultural growth has been experienced in both developed and developing countries over the years. A good example is Asia, a country that has transformed its agricultural practices by doubling its production of rice and wheat from the period of 1970–1995 by expanding agricultural areas and also adopting the use of revolutionary green technologies. This led to the doubling of rural incomes per capita and a decline in poverty. These changes were also accompanied with negative consequences to the environment which resulted from the use of inputs that were applied inappropriately for example fertilizers, crop protection products and through irrigation (FAO, 2000).

As the world population grows, the world also faces the challenge of providing food to its inhabitants as a result of increased demand. With the large and growing population, expansion of agricultural land is not a possible choice and the next solution is to increase soil of arable land productivity. Increased productivity should be met with the use of technologies that are friendly to the environment and also safe to farmers and the consumers. Innovative agricultural solutions incorporate regeneration processes that are natural such as fixation of nitrogen, nutrient recycling, maintenance of soil fertility and the structure of soil. The technologies should also protection of natural predators of insects, pests, weed, and diseases incorporating them into agricultural practices. These agricultural approaches make use of knowledge that is indigenous possessed by farmers and this can be combined with technologies that are new and based on science to provide optimum results (FAO, 2000).

6.4.2. Characteristics of Soil Conservation Technologies

Traditional farming practices mostly involve intensive tilling of soil. This was many at times associated with an increase in soil fertility as a result of mineralization of nutrients. With time farmers realized that there were long term implications such as the loss of organic matter which exposed agricultural land to both wind or soil erosion. This with time reduces the productivity of soil. To compensate for the loss of regenerative processes that are natural, farmers relied on the use of chemical inputs in order to increase productivity. Over time, the negative effects of this form of tillage have been recognized. Effective solutions to the long term impacts of intensive farming were discovered after the development of suitable herbicides. This also led to different parties such as research institutions, enterprise farmers and private companies to develop and establish farming techniques that eliminated and minimized tillage operations. Techniques were developed to suit various crops and environments (Denboba, 2005).

Some of which include zero tillage, minimum tillage, mulch till in which seeds for planting are sown into land post-harvest with little or no immediate tillage. Other practices that were later established and termed as conservative included crop rotation and use of permanent and semi-permanent ground cover and they are both commonly and currently referred to as conservation tillage. The techniques or technologies have been proven to provide stability to small, medium, and large farms.

The practice of conservation tillage protects the upper layer of soil from erosion and loss of moisture in the ground. It also leads to the improvement of biodiversity in soil by providing soil with bacteria and fungi that are beneficial. The soil organisms help with decomposition of mulch and introduce humus to soil. This stabilizes the soil structure and improves soil fertility. Previously, soil tillage practiced along with burning of plant residues in order to control weeds, pests, and diseases. In the adoption of conservation tillage, herbicides can be used for weeds that are present in farms prior to sowing. This creates the need for the use of integrated pest management technologies and systems. These measures enhance the resilience of agricultural production systems hence do not act as a hindrance of the reestablishment of balance between pests and beneficial organisms and between crops and weeds. The key elements applied in soil conservation technologies include minimum soil disturbance with the help of herbicides where necessary, mulch maintenance in a consistent manner with the use of cover crops which are either live or dead crop residues, crop rotations that are sound and conservation tillage practices based on needs (FAO, 2000).

6.4.3. Categories and Types of Soil Conservation Measures

A variety of soil conservation measures are well known and also adopted by farmers. The measures can be divided according to either their use or type. Conservation measures can be used to reduce soil erosion, land management practices such as tillage and cropping practices, they affect the problem of soil erosion and provide solutions. When rotating crops and changing tillage practices are not enough to maintain soil health and quality, a combination of measures can be employed to conserve soil.

The conservation measures and technologies that can be employed vary depending on the resources and capabilities of the farmer. They include contour plowing, strip cropping, or terracing just to mention a few. The conservation measures can be categorized into four.

- Agronomic measures-they include the use of plants or soil covers, conservation farming methods, contour farming.
- Vegetative/biological measures-they include planting barriers or vegetative strips, live fences, and windbreaks.
- Structural/physical measures such as fanyajuus, terraces, banks, bunds, cut off drains and barriers.
- Overall management such as area closures and selective clearing.

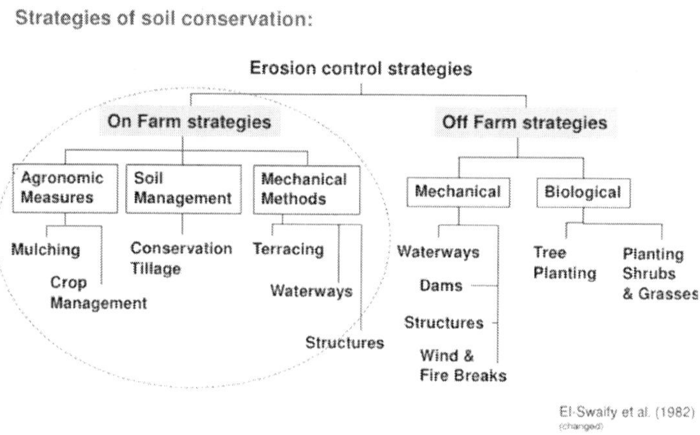

Figure 6.2: Measures and options for soil conservation.

Source: https://www.slideshare.net/IITA-CO/soil-conservation-options-in-the-savanna-of-west-africa-new-approaches-to-assess-their-potential.

When controlling soil erosion there are two approaches that are used. This entails the use of barriers or soil covers. The barrier approach uses walls or banks as earthen structures, grass strips or hedgerows to control surface run offs, reduce the velocity of wind and the movement of soil. The use of soil covers lessens rain impact and that of run offs and also reduces the amount of soil that is blown or carried away. This is done by using cover crops, mulch, minimum tillage or using agroforestry (Denboba, 2005) (Figure 6.2).

6.4.4. Factors Influencing the Adoption of Soil Conservation Technologies

Land degradation in the form of soil erosion possess a threat to food security and the sustainability of agricultural production globally. In response to this problem, governments, and supporting agencies have invested resources to conserve soil. Studies have been done in different countries to determine the factors that determine the adoption of different soil conservation technologies. Sometimes efforts to conserve soil fail to meet the objectives that have been anticipated. Some of the common factors that affect the adoption of soil conservation technologies age of the farmer, perception of the farmer on soil erosion and conservation, size of the farm, livestock holding, level of land security felt by the farmer, availability of labor, contact of farmers with development agencies, the farmers access to training and credit facilities (Mati and Mbura, 2006).

When there is land security a farmer is assured that they have control over decisions pertaining to the land and benefiting from the adoption of the soil conservation techniques. The productive and economic value of land will also determine whether a farmer will adopt conservation technologies. Another factor is whether the farmer has enough time of resources and labor investment to learn a technique and adopt it. There is also the factor or question of whether farmers have the willingness to learn and adopt the conservation technologies (Hellin, 2016).

6.4.5. Agronomic Soil Conservation Measures

They function to reduce impact of raindrops through interception hence reducing soil erosion. They also increase rates of infiltration. They also function to improve the content of organic matter in soil.

6.4.5.1 Conservation Agriculture

There are three major principles of conservation agriculture. They include minimal disturbance of soil, soil cover that is permanent and rotating crops. Practicing of soil conservation agriculture leads to the improvement of organic matter in soil after applying the technology for a period of time. Soil organic matter is important in the improvement of soil fertility and quality. They soil properties such as infiltration, water holding capacity and structure of soil (Hellin, 2016).

Keeping soil covered is the fundamental principle of conservation agriculture. After harvesting, crop residues are left on the soil surface to protect the soil. If the gap between harvesting and planting takes too long, then additional cover crops may be required. Cover crops improve the stability of the conservation agricultural system. It not only improves soil properties, but it also improves the capacity of soil to promote an increase in biodiversity in the agricultural ecosystem. The benefits of cover crops include:

- Stabilize soil moisture and temperature;
- It protects soil during fallow periods;
- Mobilization and recycling of nutrients;

- It also leads to the improvement of the structure of the soil and it also breaks compacted soil layers and hard pans;
- It can also be used in the control of weeds and pests; and
- Production of additional soil organic matter and the improvement of soil structure.

Figure 6.3: Conservation tillage.

Source: https://www.infonet-biovision.org/EnvironmentalHealth/Introduction-soil-conservation-measures.

An example is cowpeas which is an important legume throughout the tropics and sub tropics. Its properties make it an ideal cover crop. Cow peas are drought tolerant crops and can grow in the presence of little water or rainfall. It can fix nitrogen and grow even in soils with poor quality. It also produces grains that can be consumed and used as fodder for animals since they are rich in protein. Cropping systems can be designed in such a way that the soil is almost permanently covered with plant canopy. In arable crops, careful timing of sowing and planting can help to avoid soil that is not covered from being eroded. During seasons where there is precipitation. After harvesting the main crop, a green manure crop can be sown. On slopes, crops should be grown in contour lines that is across the slopes rather than being grown vertically. Planting crops along contour lines can contribute to a reduction in the speed of surface water (Mati and Mbura, 2006) (Figure 6.3).

In conservation agriculture, the only methods that should be applied are minimal and no soil tillage. It involves crop seeding without mechanical seedbed preparation and minimal or no soil disturbance since the harvest of previous crop (Figure 6.4).

Figure 6.4: A farm left with plant remains from previous harvest to provide cover to soil.

Source: https://www.johnstonnc.com/swc/content.cfm?pageid=wisc.

No-till Farming: It involves planting of seeds into the residue of the previous crop, with no tillage between harvests. No till will leave a percentage of the farms covered in crop residue. An aspect of conservation tillage is based on the selection of machinery used. The moldboard plow which is used traditionally tends to cause soil disturbance leaving soil bare and exposed to erosion. On the other hand, chisel plows leave 30 to 50% of the soil surface covered with soil residue.

Timing is also a factor of importance especially for most soil types. It is preferable that a farm is tilled in the spring after the major erosive force of spring melt has passed. Soil should not be tilled while still wet or traveled on. This practice results in soil compaction. Apart from control of soil erosion, other advantages of conservation tillage are increased infiltration of water, addition of organic matter to soil and economic gain as a result of saving of fuel and also time for the farmer. It also leads to the enhancement of habitats of wildlife for organisms in soil, birds, and small animals like field mice and snakes (Figure 6.5).

Figure 6.5: Effects of minimum tillage of soil.

Source: http://www.fao.org/3/X8234E/x8234e0a.htm

Crop Rotation: It is an alternative to continuous monoculture. The main crops are rotated, especially cereal crops for example winter wheat or forages such as clover or alfalfa. This method also involves growing of different types of crops on agricultural land instead of growing a single type. This method is employed such that soil is not used for only one set of nutrients. Growing of the same type of crop on the same land over and over depletes certain soil nutrients. Practicing soil erosion can prevent leaching of soil nutrients with the use of a single crop, it encourages the growing of an alternative crop in the following season that returns the nutrients lost by the previous crops as it draws a different ratio of nutrients. This method provides different benefits as it reduces the problem of pest and diseases thus reducing the dependency on using pesticides. Crop rotation does not give pests the opportunity to establish large numbers that can cause damage. Forage crops are used as green fertilizers on farms. They have nitrogen-fixing ability which is useful to other plants. The green fertilizers also improve the organic content of soil. All these characteristics protect soil against erosion by either wind or water. Other benefits include an increase in biodiversity, risk management and carbon sequestration. There are several factors that are to be considered if crop rotation is to be practiced this include market demand of the choice of crops, availability of labor supply, climate, the type of soil and the growing practices among others (FAO, 2000) (Figure 6.6).

Figure 6.6: An example of a crop rotation cycle.

Source: aridagriculture.com/2018/03/01/crop-rotation-agriculture/.

6.4.5.2. Contour Farming

Contour farming involves plowing, planting, and weeding along contour lines, for example, across a slope rather than up and down the slope. Contour lines are lines that run across a slope instead of up or down such that the line stays at a similar height and does not run uphill and further apart on gentle parts of the slope. Studies done show that contour farming can solely reduce soil erosion by approximately 50% on slopes that are moderate. For slopes steeper than 10%, other measures can be applied in addition to contour farming to enhance their effectiveness. If the establishment of contour lines is not done effectively, it can increase the risk of soil erosion (Figure 6.7).

Figure 6.7: Contour farming on an almost plain agricultural field.

Source: https://www.johnstonnc.com/swc/content.cfm?pageid=wisc.

Figure 6.8: Contour farming practiced on steep slopes.

Source: https://www.nrcs.usda.gov/wps/portal/nrcs/detail/ga/programs/planning/wpfp/?cid=nrcs144p2_021804.

Contour ridges are mainly used in areas that are semi-arid to water that is harvested and in higher areas for potato growing. Trash lines are created by laying crop residues or trash in lines along the contours. Growing of crops along contour lines slow down surface runoff and trap soil that is eroded, eventually farming terraces. The contour lines can be destroyed by termites that can feed on trash placed along the lines. Grass barrier strips are planted along contours. They are planted with fodder grass such as Napier or can be left with grass that is natural. Using grass barriers are measures of soil conservation that are effective in the quick absorption of water and on steep slopes as 30%. A farmer or individual may want to mark contour lines in order to decide where to plant barrier strips or lay trash lines. One should begin by first studying the lines and visualizing where the contour lines will run. This can best be done by one person directing another in walking to the opposite side of the area that needs to be contoured such that they stay at the same height as the first person. A person can use A-frames which are easy to make. They can be made by material that is readily available and be easily used. It consists of three pieces of wood assembled to make a shape of a capitalized letter A. the A frame is held in an upright position and a weight on a string hangs down from the top of the A and it acts as a plumb line. If the A frame is on a perfect level ground, the string crosses the A horizontal bar at a designated point. The point is marked during the calibration (FAO, 2000) (Figure 6.8).

To use the A frame, it is moved across the slope, making sure its legs are level at each point by checking if the string crosses the bar that is horizontally placed at the point of calibration. Another method would be to move the front leg until the string shows the frame is level. The position of the legs on the ground can be marked with the use of pegs, and the frame is pivoted around to mark a new point on the slope (Figure 6.9).

Requirements for making an A-frame:

- Straight poles that are wooden (2) 1.5–2 meters long;
- String, hammer, nails, and a pencil; and
- A stone that is round.

Step 1: Making the A-frame.

- Use the nails and poles to make an A shaped frame;
- At the top of the A, tie a single end of the string; and
- On the other end, tie a stone so that it hangs beneath the horizontal crossbar of the A.

Figure 6.9: Making an A frame for creating contour lines.

Source: www.iirr.org.

Calibration of the A-frame.

- Place the A frame in an upright position and on a level ground where the two legs stand;
- Holding the A frame, use the pencil to lightly mark the crossbar where the string crosses it;
- Turning the A-frame round, so that each leg exactly stands where the other had been placed; and
- The two marks should be close proximity. The center shows where the string should cross in the A frame was on a ground was level. Make a heavy pencil mark on the bar with a knife at this point (Figure 6.10).

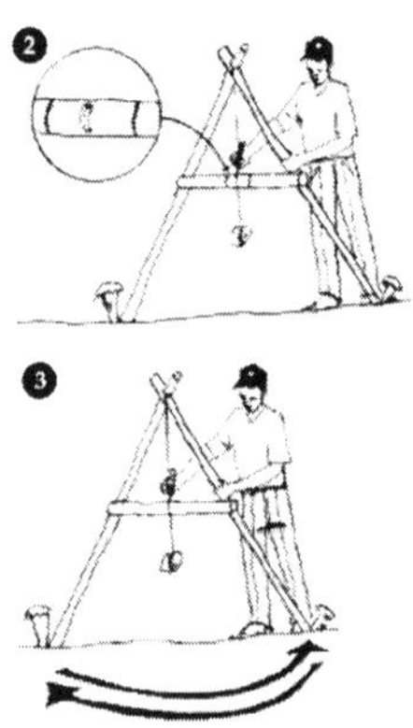

Figure 6.10: Calibration of the A frame.

Source: www.iirr.org.

Step 2: Locating the contour lines using the A-frame.

- On the slope, chose a place to begin. Put the frame in an upright position and place a mark where the first leg stand using a stone or a visible marker.
- Without moving the first leg and the frame in a position that is up right, move the second leg up or down the slope until the string crosses the crossbar exactly at the heavy pencil mark.
- Using a suitable marker, mark where the second leg stands.
- Keeping the second leg in a similar position, lift the first leg up and pivot it around. Move the frame up and down the slope until you find a position where the string crosses the cross bar at the pencil mark that is heavy.
- Using a suitable marker, mark the current position of the first leg.
- In a similar manner, continue following the same steps as above till you reach the end of the field.
- A contour line will be marked with the stone or other marks, they will all be at a similar height on the slope. The markers are usually in straight line and if necessary a curve can be made by moving them a little up or down.
- To mark another contour line, move up and down the slope at a determined distance for example 20 meters on a slop which is gentle or a 1.5 meters drop on slopes that are steep and repeat the same process as above (Figure 6.11).

Figure 6.11: Locating contour lines using the A frame.

Source: www.iirr.org.

Step 3: Preparation of the contour lines.

- After making the contour lines, prepare the contour lines until they are ready for planting. The width of each area that is to be prepared should be one meter. They serve as a guide in the event of planting.

Step 4: Plant seeds of nitrogen-fixing trees and shrubs.

- Two furrows are laid out on each contour line prepared. Two or three seeds can be planted at a hill at a distance of about 0.6cm. The seeds should be covered with soil. When the plants are grown fully, the hedge rows bank the soil and serve as a source of fertilizer. Examples of species include *Desmodium* and *Leucaena* species.

6.4.5.3. Cover Crops/Mulching

They are crops planted to reduce wind and water impact on bare soil. The cover crop absorb the rain impact, reduce the speed of run offs, hold soil in place and increase water infiltration of soil and as a result there is less run off. Some of the cover crops include alfalfa, rye, sweet clover and winter wheat. They are planted in areas like steep slopes which are susceptible to soil erosion also in streams, river banks and grassed water ways as well as around wells to protect ground water from contamination. During winter, winter wheat and rye are planted as cover crops to control soil erosion during spring runoff (Figure 6.12).

Figure 6.12: The use of dead crops as soil cover/mulch.

Source: https://wjfarm.wordpress.com/2010/07/31/straw-vs-hay-mulch-happens/.

Intercropping-involves planting of different plant species in a farm field for example legumes can be planted together with rows of soybean. The technique can be used by farmers who don't have the economic capability to take his or her entire crop of corn out of production (Figure 6.13).

Figure 6.13: Intercropping.

Source: Will Critchley, Amsterdam, the Netherlands (Reproduced from WOCAT, www.wocat.net).

6.4.5.4. Strip Cropping

Involves alternating strips of small grain or forage crops with row crops like corn. It is used as a measure of control for soil erosion by decreasing wind and water velocity. The rows of cereal and forage tend to trap sediments that may be channeled into watercourses or sources. It is the most effective means of controlling soil erosion on slopes when it is placed along land with contours. In the control wind erosion, it best works if the strips are placed at right angles to the direction of winds that are prevailing. Another benefit is that organic matter is added as a result of decomposition of the forage or cereal crops rows (Figure 6.14).

Figure 6.14: Strip farming in Wisconsin, 1957.

Source: https://en.wikipedia.org/wiki/Strip_farming.

6.4.5.5. Agroforestry

It describes land use systems where crops are grown together with agricultural crops, pasture or livestock. Their interactions are usually both ecological and economic between components of the system. It has been practiced for years. It focuses on a wide range of working trees grown on farms and in landscapes found in rural areas. Some of the trees include fertilizer trees that are responsible for the regeneration of land, contribute to the health of soil and food security. They also contribute to nutrition; fodder trees improve livestock production for small scale farmers; timber and fuel wood trees provide shelter and energy; some trees provide medicinal resources that help to combat diseases; some trees also produce gum, raisins, and latex products. Most of the trees serve multiple purposes and provide a wide range of benefits (Denboba, 2005) (Figure 6.15). Some of the benefits that have been proven include:

- Poverty reduction through the increased production of products for human consumption or sale;
- It contributes to food security by restoring the fertility of soil for food crops, fruit production, nuts, and edible oils;
- If support is provided, it ensures fair deals for farmers who are female and farmers from rural disadvantaged areas whose rights to land are not secured;
- It reduces incidences of deforestation and pressure exerted on woodlands by the provision of wood to be used as fuel from

farms;

- It increases the diversity of farm tree crops and tree covers that protect farmers from the effects of global climate change;
- It improves nutrition which reduces the impact of hunger and illnesses that hare chronic such as HIV/AIDs; and
- It augments the access to trees with medicinal resources which can provide medication to a large percentage of the African population.

Figure 6.15: *Sorghum grown under Faidherbia albida and Borassus akeassii near Banfora, Burkina Faso.*

Source: https://www.infonet-biovision.org/EnvironmentalHealth/Agroforestry.

6.4.5.6. Use of Natural Fertilizers

Natural fertilizers include the use of manure from livestock, mulch, municipal sludge and legume plants such as alfalfa or clover. Manure and sludge can be spread over land and then can work its way into soil. Timing of application is based on following strict guidelines since manure and sludge can be contamination sources to water if they are not properly handled. The legumes that are grown can also decompose and for manure which also referred to as green fertilizer. Natural fertilizers improve and increase nutrients that are essential for plant growth such as nitrogen, phosphorous, and potassium. They have the benefit of adding organic matter to soil.

6.4.5.7. Integrated Pest Management

Uses different techniques designed to prevent or cut the use of pesticides that have a chemical nature hence reducing risks to the environment. Crop rotation is the backbone of integrated pest management. Practicing crop rotation annually starves pests and therefore reduces their numbers. It is therefore a proven method of controlling pests. Integrated pest management encourages the use of crops that are pest resistant and using biological controls such as using pest predators or parasite to control the population of pests. This method take time but ensures environmental safety.

6.4.5.8. Management of Soil Salinity

Soil salinity is caused by irrigation using water that is salty. After evaporation of this water salt is produced and left behind. Soil can lead to the breakdown of the structure of the soil and this can affect the productivity of land by causing soil infertility and a reduction in plant growth. Some of the ions responsible for soil salinization include sodium, potassium, calcium, and chlorine ions. Salinity of soil affects metabolism in crops and also erosion. Salinity can occur as a result of over irrigation in areas that are dry or in areas with saline water tables that are shallow. Salts are deposited on the upper layers of soil as resultant products of infiltration. Irrigation only functions to increase the rate of deposition of salts. In 1970 the best known shallow saline water capillary action took place in Egypt after the Aswan dam was constructed which resulted into concentrations of salts that were high in the water table and this resulted to soil salinity. Excess salination can be prevent by humic acid in the event of excessive irrigation. Both anions and cations can be fixed and eliminated from root zones by humic acid. Lowering of water tables can be done by planting crops that are tolerant to conditions that are saline thus reducing the capillarity rate and enrichment by evaporation of salts on the soil surface. Some of the crops that are tolerant to salt bush (Figure 6.16).

Figure 6.16: Salt deposits on the former bed of the Aral Sea.

Source: https://en.wikipedia.org/wiki/Soil_conservation.

6.4.5.9. Soil Mineralization

For plants to realize their phytonutrient potential, active mineralization of soil is sometimes done. The process involves the addition of crushed rock or chemical supplements to soil. The purpose of doing this is to deal with the loss of minerals from soil. A number of minerals can be used some of them include substances like phosphorous, zinc, and sometimes selenium. Floods also can be a source of minerals if they do not pose a big threat to humans and the environment. It can lead to addition of minerals especially to flood plains and this rejuvenates the soil chemistry in relation to minerals.

6.4.5.10. Soil Organisms

Soil organisms such as worms excrete fecal matter in the form of casts. This provides a balanced selection of minerals and plant nutrients to soil into a farm making them accessible for root uptake. Casts from earthworms are five times richer in available nitrogen, it is seven times richer in phosphates and eleven times richer in available potash than the surrounding upper 150 millimeters of soil. Earthworms can generate up to 4.5 kilograms of cast per year. Earthworms also increase the porosity of soil and also creates channels that enhance soil aeration and soil drainage. Other organisms that are important to soil include nematodes, mycorrhiza, and bacteria. Soil that is degraded requires the use of fertilizers that are synthetic in order to produce high agricultural yields. The lack of a good soil structure increases incidences of soil erosion and it also carries nitrogen and other pollutant to rivers and streams. Increase in the organic matter of soil increases the water holding capacity of soil (Figures 6.17, 6.18).

Figure 6.17: Yellow fungus, a mushroom that assists in organic decay.

Source: https://en.wikipedia.org/wiki/Soil_conservation.

Figure 6.18: Fecal matter of earthworms in the form of casts.

Source: https://en.wikipedia.org/wiki/Earthworm#Environmental_impacts.

6.4.6. Vegetative/Biological Measures

Biological conservation measures work to protect soil using the vegetative cover which is usually dense. It prevents splash erosion, reduces the speed of surface run off at the same time increase infiltration, increases roughness of the soil surface hence reducing infiltration and runoff and it also stabilizes soil aggregates using its roots and organic matter. They are one of the most effective ways to be used in the control of soil erosion since they do not require a lot of resources.

6.4.6.1. Wind Breaks/Shelter Belts

Windbreaks can sometimes be referred to as shelterbelt. It is a vegetative barrier that is created to facilitate the reduction or elimination of wind velocity hence reducing wind erosion. Windbreaks can be distinguished from shelterbelts. Wind breaks possess one to five tree rows or shrubs while shelter belts have six or more rows wide. Windbreaks are planted on different boundaries of a field but generally on the west, southwest boundaries in order to protect the fields from prevailing winds. Protection of soil is maximized downwind of the barrier, where the wind speed is significantly reduced for a distance of 15 to 20 times the height of the trees. The windward side of the break receives protection for three to five times the total height. The benefits are far beyond the control of soil erosion. It improves agricultural produce and its quality, improves the performance of livestock as a result of reduced abrasion from soil being blown. The rows of shrubs and trees increase the aesthetic value of land significantly hence increasing the value of land. In countries that receive snow, it traps the snow and thus provides increased moisture content during the crop planting season. It also improves the habitat of wildlife since they provide places where animals can seek shelter from their predators and also protect themselves from weather conditions considered as harsh (FAO, 2000) (Figure 6.19).

Figure 6.19: Wind breaks or shelter belts.

Source: https://www.nrcs.usda.gov/wps/portal/nrcs/detail/ar/home/?cid=nrcs144p2_027201.

6.4.6.2. Grassed Waterways

Grassed waterways are permanently vegetated saucer-shaped channels made to carry surface runoff across land without causing erosion. It is employed in areas that experience gully or rill erosion to overland flow of water that is concentrated. This conservation measure prevents soil erosion by slowing the flow of water. The flowing water is the safely carried to an outlet that is stable such as a stream or drainage ditch. The advantage of employing this method is that it does not alter the course of water and once they are well established they can be crossed by farms, construction or forestry vehicles (Figure 6.20).

Figure 6.20: An area with a grassed water way.

Source: https://www.johnstonnc.com/swc/content.cfm?pageid=wisc.

6.4.6.3. Vegetative Strips/Cover

There are various ways in which grasses or trees can be used as vegetative strips or cover. They can be planted to provide catchment to soil, excess nutrients and chemical pesticides that move over land before their entrance to water sources. They lead to the formation of bunds or terraces due to erosion that results from the tillage of land. This is the movement of soil down slope when land is being cultivated. The effect of vegetative cover that is spread is multiple including an increase in ground cover, improvement of soil structure and infiltration as well as decreasing erosion by wind and water.

6.4.6.4. Reforestation/Afforestation

It areas where land is highly degraded, it is important to regenerate the balance of soil and water. Trees perform various functions such as protecting soil from erosion and it also functions to preserve moisture in soil. Trees function as windbreaks and can be planted in areas where they were nonexistent. They reduce the velocity of wind. They are planted across the area at a ninety-degree angle to prevailing wind. Trees check the spread of deserts or sand dunes and also prevent the blowing away of top fertile soil that is useful for agriculture.

6.4.7. physical/Structural Measures

They are structures built to conserve soil. They aim at slowing runoffs allowing it to infiltrate into the soil. They reduce the speed of surface runoffs and offer protection against excessive damage.

6.4.7.1. Drains, Ditches, Broad Beds and Furrow Terraces

Cutoff Drains: They are dug across a slope to intercept surface runoffs and carry it safely to an outlet such as a canal or a stream. The drains protect land that is cultivated, roads, and compounds from runoffs that are controlled and this diverts water from gully heads.

Retention Dishes: Retention ditches are dug along contours. The ditches catch and retain surface run off and holds the water until it infiltrates into the ground. The ditches can be used to harvest water in areas that are semi-arid.

Infiltration Ditches: Surface runoffs can be harvested from roads and other water sources. They consist of ditches that are 0.7–1.5 meters deep. The ditches are dug along the contour, upslope from a crop field. Water is diverted from the roadside into the ditch, which is blocked from the other end. The trapped water then infiltrates into the soil. For soils that contain hardpans, the water does not infiltrate directly into the soil. Instead, it moves downslope just below the surface, towards the crops.

Broad beds and Furrows: In this system, the runoff water is diverted into field furrows that can be 30cm wide and deep. The field furrows are blocked at the lower end. When a furrow is completely filled with water, the water backs up into the furrow head and flows into the furrow in the next field. There are broad beds in between the field furrows about 170 cm wide where the crops will be grown.

6.4.7.2. Terraces

A terrace is an earthen ridge that is constructed within a water channel along the upper side. There are several design options for terraces but the most common is the ridge and channel which are permanently grassed. Terraces are designed to absorb run off on a slope and reduce erosive action on slopes. Water speed in channels is reduced and moves along a vegetative channel to a safe, outlet that is stable such as a waterway that is grassed or drops inlet.

Terraces-Fanyajuu: They are terraces constructed by digging ditches and afterward heaping the soil. The results are the forming of bunds in the upper sides of the ditches. A small ledge prevents soil from sliding back and it exists between the ditch and the bund (Figure 6.21).

Figure 6.21: *Fanyajuu terraces in a semi-arid area which have developed over time into benches. Note the well-established grass strips along the bunds.*

Sources: https://www.accessagriculture.org/category/147/cross-slope-barriers

Terraces-Fanyachini: It is similar to fanyajuu, the difference is that it is placed on the lower side of the contour trench, not on the side that is up the slope. It conserves the soil and diverts the water. The embankment that is resulting can be used to cultivate fodder. They are easy to make in comparison to fanyajuu but do not lead to the formation of bench terraces as time passes. They can be used on slopes up to 35%.

6.4.7.3. Drop Inlets and Rock Chutes

A drop inlet consists of a vertical intake pipe and a horizontal underground pipe. Water enters the vertical pipe at the surface of the ground, and falls below where it is guided safely through concrete metal or plastic pipes that are large into a steam or large. Rock chutes are rock piles that are designed to move water that is concentrated in areas and flows over steep slopes. These techniques are both used where there are rapid changes in elevation thereby protecting soil from erosion (Figure 6.22).

Figure 6.22: A rock chute.

Source: https://www.johnstonnc.com/swc/content.cfm?pageid=wisc.

6.4.7.4. Bank Stabilization

Involves any measure that can hold soil in place on the bank or course of water. Soil can be eroded by stream current, waves, ice, and surface runoff. Bank stabilization reduces soil erosion, leads to better water quality and improves the aesthetic value of the environment. Three methods can be used in the control of stream bank erosion they include rip rap, gabion baskets and re-vegetation. The first two methods use rocks that are loose to break stream water impact on the bank and also protect loose soil surface beneath it. An advantage of rip rap is that it does not give to the pressure of ice and frost contrary to concrete that may break. Gabion basks are wire basket that are filled with rocks and are commonly used in slopes that are steep or where water movement is at a high speed. Shoreline planting can be used to stabilize stream banks. The movement of soil over water can be slowed down by natural grasses, shrubs, and trees. They also trap sediments hence preventing them from reaching water sources. Bushes will also provide an excellent habitat for wildlife.

6.4.7.5. Sediment Control

Conservation efforts can be made to prevent soil sediments or silt from reaching water sources. A silt fence can be constructed to filter surface runoff and trap sediments using cloth filters. The constructed structure can also function to reduce the occurrence of soil erosion by decreasing the speed of wind. Sediment traps contain stones and in addition may have a filter cloth which together prevents the movement of soil particles and slows their velocity. The filter cloth functions to prevent fine particle from entering the storm sewer. A sediment pond allows the temporary storage of water from runoff which is afterward guided to an appropriate outlet to a stream or ditch. These sediment controls as measures of soil conservation will only continue to function properly if they remain well maintained. After sediments are trapped by these controls they should be removed properly and stabilized and these will allow the effective function of the controls.

6.4.8. Regeneration Measures: Change in the Use of Land, Area Closure and Grazing

Management measures are often applied to land for grazing in instances where other measures simply don't work without a fundamental change in the management of land. Protection from grazing, allows vegetation cover to regenerate. Certain measures are important for the rehabilitation of areas that a degraded badly and in areas where technical measures and interventions are not adequate on their own. In some areas where there is intense grazing, fodder crops are grown and used to feed livestock. One of the major advantages of conservation measures often do not require high investments such as labor or money. Leaving land bare or not using it can create pressure on neighboring land which can also be already in poor condition and this increases its vulnerability to further degradation (Mati and Mbura, 2006) (Figure 6.23).

Figure 6.23: Rehabilitation of degraded rangeland in its initial stage: Stone lines are established after the area has been cleared of invasive tree species (South Africa).

Source: https://www.infonet-biovision.org/EnvironmentalHealth/Introduction-soil-conservation-measures.

6.5 BENEFITS OF SOIL CONSERVATION TECHNOLOGIES

6.5.1. Economic Benefits

One of the main reasons farmers should consider conservation technologies, is the economic benefits the technologies bear. The profits made is an annual concern and is based on influenced by production costs. One of the main reasons for adopting conservation technologies by farmers is an increase in land and labor productivity, lower cost of operations which has led to increased income for farmers. Studies done have indicated long term substantial gains especially to production. For examples in the United States the application of the no till technique in maize production was 90% greater than maize that has been tilled even in the first years of adoption and experienced an increase of a 100% after 20 years of practice.

6.5.2. Risk Management

Improving the quality of soil leads long term benefits such as gains in productivity and production that is stable. This reduces the risk to farmers by stabilizing their yields especially in drylands and areas experiencing low precipitation.

6.5.3. Environmental Impact

Conservation tillage has been practiced in many areas such as the United States, Brazil, Australia, and Canada among others. Considerable information which is now available has shown that they are environmental benefits from adopting the said practices some of which even have global impacts. Technologies such as conservation tillage lead to the improvement of the quality of soil, reduce soil erosion and slow surface run-offs from farmland. It also reduces the carbon emissions released into the atmosphere. The above-mentioned benefits help countries to meet the stringent and sometimes demanding regulations for environmental protection.

6.5.4. Soil Quality

As discussed previously, it is one of the soil characteristics or properties which is largely determined by the amount of organic matter that is present in soil and it also responds to management practices. The practice of conventional agricultural methods over time lead to a reduction in the content of organic matter in soil. For example, a survey done in the United Kingdom by the soil survey and research Centre from 1980–1995 shows a significant decrease in the number of sites with a high organic matter. Contrary to these as a study done Ontario, Canada shows an increase in organic matter as a result of adopting conservative tillage techniques after stopping the use of conventional tillage techniques which makes use of disc or moldboard plowing.

6.5.5. Soil Erosion

It is an environmental threat that is experienced worldwide. A large part of arable land in the world has been left bare as a result of soil erosion. Soil quality continues to be lost at a rate of 10 million hectares yearly. This trended has been largely contributed by the intensification of agricultural methods especially the use of conventional methods and adopting cropping systems that leave surfaces bear during rainy seasons. Agricultural produce from eroded soils tends to be lower than those from good quality soils or protected soils by up to 35%. the erosion of soil increases the cost of agricultural production by at least 25% yearly. The use of crop residues and reducing tillage can be adopted to reduce the soil erosion process. For example, in the United States adopting the use of conservation tillage helped to reduce erosion in farms by 30% and 31% in 1997, which was equivalent to saving 1 billion tons of soil annually.

6.5.6. Water Quality

Soil erosion affects the quality of water resources as most of the surface runoff ends up in water sources. It leads to the sedimentation of water sources and also contamination by agricultural inputs residues from pesticides and also some fertilizers which render water from these water sources unfit for human consumption. High levels of sedimentation affect the aquatic ecosystem by destroying habitats meant for aquatic life. It also decreases the storage capacity for water reservoirs. This as a result increases the need for treating water and dredging which has extra cost implications for parties involved.

6.5.7. Reduced Carbon IV Oxide Emissions and Improved Quality of Air

Maintenance of mulch in the practice of conservation tillage leads to an increased ability of the soil to sequester carbon IV oxide and this leads to a reduction in its emissions hence protecting the atmosphere. These tillage mechanisms also increase the organic content matter in soil which improves nutrients in soils and the water holding capacity of soil. Tillage has been proven to increase the oxidation of organic matter which then produces carbon IV oxide in return increasing atmospheric circulation of this greenhouse gas. Conservation tillage can reduce the emission of this gas by 80%. Carbon can be sequestered by adopting conservative tillage techniques. According to a lot of research done over the years, it is evident that the burning of fossil fuels releases carbon IV oxide and similarly, research has shown that soil tillage can also contribute to a percentage of the gases released to the environment which consequently leads to global warming which is a global environmental problem. The use of conservation tillage techniques significantly reduces the amount of fuel to be used by farming equipment such as tractors across fields or the amount of trips taken. The reduction in the use of fuels is dependent upon the soil structure, moisture content and the tilling system in place. The less the amount fuel use the less the carbon emissions released to the environment. By reducing soil erosion especially by wind, conservation tillage reduces the amount of dusts that is released to the atmosphere which is a source of pollution in many parts of the world (FAO, 2000).

6.5.8. Wildlife Habitat and Biodiversity

Soil conservation practices that incorporate leaving 70% crop residue to cover the soil, provides a favorable environment for organisms that inhabit the soil, different bird species, small mammals and reptiles. Similarly maintaining tree and other vegetation on agricultural land creates an improved quality of habitats especially for birds. Organisms present in soil like earthworms, bacteria, and fungi receive the most benefits from conservation technologies and techniques and they in turn help by improving the quality of soil by improving its structure and fertility (FAO, 2000).

Increased adoption of conservation technologies leads to increased production in areas where agriculture is practiced. This conservation technologies further increase opportunities to improve natural resource management and broadens areas for conservation of natural ecosystems and biodiversity. Currently, there is no sufficient evidence to support the change in pollution of wildlife species in relation to adoption to conservation technologies, however existing evidence points to the fact that farms that produce high yields allow more species to persist. If a farm produces high yields in their existing farms, they are less likely to have the pressure to clear land that is intact and is inhabited by wildlife. Active involvement of community members allows for knowledge sharing and also provides a mechanism that is effective to ensure sustainability that is long-term (FAO, 2000).

6.5.9. Social Impact of Adopting Soil Conservation Technologies

In industrialized and developing countries the need for sustainable management practices for sustainable development presents challenges that differ in nature. Soil conservation technologies increase productivity of land. Agriculture in industrialized countries finds ways of responding to environmental and food quality issues of concern. Policies and regulations in developing countries are now increasingly requiring farmers to adopt soil conservation technologies that act to minimize the impact of technology that is modern on natural resources, restoration of ecosystems and also protect the environment, preserve biodiversity on the farms as well as non-farm situations. The use of crop residues conservation technologies will help to address these issues and concern at the same time optimizing the potential of land in support of sustainable agricultural production. However, for farmers to adopt these technologies in many countries they need to be convinced that these new technologies are better in both short and long term instances.

For developing countries, there is the need to reduce poverty and reversal of other social consequences of productivity that is low and the deterioration of sources of natural resources. In some countries, in order for farmers to the meet increased demand for food by expanding agriculture to areas that are marginal and they also encroach forest areas. The search for high productivity of land has led to the excessive use of agricultural inputs and natural resources overexploitation especially land and water resources. The adoption of soil conservation technologies has the potential to make agriculture more efficient by leading to a reduction in requirements of labor needed for the preparation of land for farming, farming, and removal of weed at the same time conserving water and land. Since in some cases farm operations are still done manually, conservation tillage reduces drudgery and permits the release of labor for other economic and social needs (Denboba, 2005).

CHAPTER 7

Flood Prevention Technologies

CONTENTS

7.1. INTRODUCTION

Floods have existed since the first day of existence of the entire globe. We read of them in the Bible, the stories of Noah, where a large population was brought to extinction by floods. We also hear stories of how floods have had impacts on the entire globe. The globe's history does have a lot to do with floods. The word flood is not really a new word to us. Basically, floods mean the overflow of water on the ground once the underground level has been saturated with water. Floods can be experienced anywhere in our surroundings. Lowlands are the most affected areas by floods. But we can talk of river banks, once the river is quite filled with huge amounts of water, some of the water tends to overflow for the river no longer has enough room for its accommodation. Riverbanks end up being covered with water. Lakes may also flood. Oceans and seas may flood more so in periods of high tides when the sea level has risen above its normal positioning. Floods are of much negative impact to the environment more than they are of positive of impact. They lead to distraction of lands, distraction of property, distraction of normal lifestyle of people, extinction of objects and living things among other negative impacts.

7.2. FORMATION OF FLOODS

Floods are formed when the duration of infiltration is quite minimal as compared to the amount of water which needs to be infiltrated underground. Most of such situations occur when heavy rains are experienced on the earth's surface within a brief period. Such rains do not give the ground enough time to absorb the water they have released hence the chances of flooding a quite high. Flooding may also be formed when the glacier melts at a higher rate than normal giving the ground limited time to absorb the melt water. Water body levels tend to rise as huge amounts of water are fed into them, either from rain water or glacier, at a higher rate than normal leading to breaking of their banks which cause flooding in their surroundings.

7.3. TYPES OF FLOODS

Floods come in different forms. Most floods are categorized depending on the areas they mostly occur; others are categorized depending on the amount of damage they cause to the surroundings. They can also be categorized depending on the causative agents such as rise in water body levels, deforestation among other causes (Figure 7.1).

Figure 7.1: One of the many types of flooding experienced on the earth's surface is the coastal flooding.

Source: https://medialibrary.climatecentral.org/extreme-weather-toolkits/coastal-flooding.

Coastal flooding is dominant in the coastal regions of our continents. It is mainly experienced due to the rise in sea levels during high tides which are periods when there is a pull of the moon and the sun. Severe storms do also occur mostly in coastal regions leading to flooding of the surrounding areas and sometimes might pose a great threat to human life. Low lying sea sides are also known to have high rates of flooding since their level is quite below the sea's normal level hence accumulation of water in them.

The second type of flooding is the river flooding. As the name suggests river flooding occurs alongside rivers. Once the water in a river exceeds the normal capacity that can be accommodated, the strong forces of the river tend to bursts the river banks as they seek more space for their accommodations. Bursting of river banks may pose a great danger to the surrounding areas for the water that breaks out of the river is under high pressure and sweeps away any barrier it comes across as it tries to free itself (Figure 7.2).

Figure 7.2: An example of river flooding.

Source: https://www.reuters.com/article/us-usa-weather/central-u-s-faces-high-risk-of-river-flooding-even-as-rains-ease-idUSKCN1T012K.

Flash flooding is the third type of flooding commonly experienced on the earth's surface. It is mainly caused by rain water. When heavy rains fall within a short period of time they tend to suffocate the ground level. The suffocation is that the rain falls under a very short period leaving the ground with minimal amount of time to absorb the water. Flash flooding tend to subside quickly for once the rain stops, the ground takes its time to absorb as much water as it can. However, as long as the floods last, they pose a great danger to humans. This is because water from flash floods moves at higher speeds and it tends to sweep away any barrier it comes across leading to damage of property (Figure 7.3).

Figure 7.3: Flash flooding.

Source: https://www.aljazeera.com/news/2019/09/dozen-dead-flash-floods-hit-western-india-190927083502589. html.

The other type of flooding is the ground water flooding. It is not that common though once it occurs, it lasts for quite a long period of time. During the rainy seasons, the ground absorbs the water that falls from the rains. The absorption rate reduces as more water is being absorbed since the ground gets saturated. Once the ground hits its maximum saturation level, it can no longer absorb any water onto it. In such situations, water rises above the ground forming a water body like surrounding. This type of flooding last for quite some time for the water underground has to be less saturated for the water above the ground to be absorbed. It forces resettlement of people from their residents for they cannot reside in a water logged surrounding (Figure 7.4).

Figure 7.4: Ground water flooding.

Source: https://www.tobinconsultingengineers.com/news/groundwater-flooding-investigation-ireland/.

Drain and sewer flooding are not necessarily caused by weather conditions. They may occur in drought periods. Mostly caused by maybe the destruction of a drainage system which might flood the surrounding with the drainage water. Sewage systems may also break at some point flooding their surroundings. They can be said to be the most disadvantageous type of flooding for the sewage leaves the surrounding with a bad odor and other poisonous components which might cause illness to living organisms. This is because sewage systems contain waste water with so many poisonous components which need to be disposed off properly and in the right locations (Figure 7.5).

Figure 7.5: Drain and sewer flooding.

Source: https://www.rytechinc.com/blog/why-sewage-flooding-in-your-home-is-a-ticking-time-bomb/.

Catastrophic flooding is a type of flooding that occurs due to collapsing of dams. Dams are majorly constructed along rivers to reduce the flooding rates in the surroundings as the river banks bursts due to too much water in them. Dams tend to hold quite huge amounts of water in them and in situations where the wall collapses, catastrophic flooding occurs. This is because the accumulated water breaks free under high pressure which had developed over the long period of time it had been accumulated. The collapsing of the wall maybe caused by the disaster such as earthquakes or even way too much pressure to the wall by the long prevented water by the dam. Catastrophic flooding may also occur in oceans leading to catastrophe such as Tsunamis which are caused by high magnitude earthquakes experienced in sea beds. The modification of drainage systems such as rivers by natural activities such as volcanic eruptions may also lead to flooding for the water is heated under high pressure leading to its accumulation in high speeds around the surroundings of the drainage water system.

Urban flooding is a new type of flooding which has not been in existence for long. As the name suggests, it is experienced in urban areas. Most urban areas are covered with pavements which are impermeable surfaces. Due to their impermeability they prevent rain water from supping into the ground level. This leads to accumulation of more and more water above the ground which turns into floodwater. Urban flooding us quite dangerous for roads are rendered useless for they are flooded. Vehicles are washed away by floodwater. Buildings with weak foundations are destroyed, businesses are destroyed among other negative effects caused by flooding in urban centers

(Figure 7.6).

Figure 7.6: Urban flooding.

Source: https://www.upi.com/Science_News/2017/08/14/Urban-flooding-on-the-rise-as-countryside-dries-up/8471502730826/.

Estuary flooding is another common type of flooding. It occurs in river mouths where a river flows into the sea. They are mostly caused by combination of several storm surges formed by wind under the low pressure since the estuary is more of a wide opening into the sea. Estuary flooding maybe quite dangerous for storm surges may rise up to a level of ten feet above the ground causing major damages and destruction of property (Figure 7.7).

Figure 7.7: Estuary flooding.

Source: https://niwa.co.nz/publications/wa/vol11-no1-march-2003/what-happens-in-estuaries-during-floods.

7.4. CAUSES OF FLOODING

Flooding do occur all over the globe. There are various factors which lead to flooding in our surroundings. Some of the factors can either be natural factors or human factors.

One of that natural factors which a major cause of flooding is heavy rains. In our environments we are surrounded with systems designed to move rain water into its appropriate locations be it harvesting for future use or construction of ditches to help in draining the rain water. However, heavy rains end up filling our water reservoirs and the ground is saturated. Our drainage systems are overwhelmed by too much water. As the heavy rains continue, there lacks room for accumulation of more rain water which now leads to flooding.

Large rivers may lead to flooding. Rivers whose source is mountain tops maybe overwhelmed with too much water than they can bear in situations where the melting rate of ice from the mountain tops is quite high than usual. This lead to overflow of water from the river into surroundings as the water seeks a free place for its accumulation. Heavy rains do also lead to too much water in our rivers hence the bursting of river banks as more free space is being sought by the excess water in the river.

Humans have constructed dams which are basically walls build to reduce the rate of flooding in major streams. Over time such walls tend to get old hence their vulnerability increases. More pressure does also accumulate due to the water being held captive by the walls as it released to the other side in small amounts and in most cases through a narrow opening. Due to the old age, the high pressure and the increased vulnerability rate the built wall ends up breaking unexpectedly leading to the unleashing of torrents of water under very high pressure which leads to catastrophic effects to the surroundings. Such type of flooding caused by the collapsing of dams is more dangerous than the initial flooding which was being prevented through construction of dams for the water is released under very high pressures from its initial position.

Most urban centers have been surrounded by drainage system which help in draining rain water for most of the cities surface is covered with pavements which are made of concrete making them impermeable. During heavy rains, the drainage systems are overwhelmed with too much water. This lead to overflow of water from them into the impermeable surfaces of our urban areas which in most cases are low level areas. Since the surfaces are impermeable, it means that more water accumulates on the surfaces leaving them flooded for the water has nowhere to go.

Storm surges and tsunamis are unpredictable natural causes of flooding more so in coastal regions. Storm surges which are formed by heavy winds and tides in periods where the moon and the sun are being pulled by the earth's surface lead to the rise of sea water up to a level of over twelve feet above the ground. Such a rise means that all tall buildings on the offshore side are completely destroyed. Tsunamis are the most dangerous of all. They are caused by underwater earthquakes which might be catastrophic to an extend of causing volcanic eruptions on sea beds. The eruptions warm up the water increasing its pressure hence its release onto the offshore side is very catastrophic leading to massive destruction of people and property.

Water bodies whose surface is very steep are more likely to experience flooding more so in periods of heavy rains. This is because, naturally, the rate of runoff water in them is quite high. In heavy rain seasons, as more and more water accumulates in them, it means that the water body will burst its banks fast since the excess water in it is moving at a very high speed. Such flooding maybe catastrophic for the water tends to be released in a sloppy surrounding encouraging sweeping away of objects at a very high speed.

Lack of vegetation cover on the ground surface does increase the chances of flooding. Vegetation does reduce the rate at which surface runoff water flows increasing its infiltration rate into the ground level. Unavailability of vegetation means that surface runoff will flow at a very high rate with minimal chances of infiltrating into the ground. The common type of flooding experienced due to lack of vegetation cover is the flash flooding. This is because as heavy rains fall within a brief moment more so after a very long period of drought, the water flows on the ground at a higher rate with minimal chances of infiltrating increasing its accumulation rate above the ground leading to flooding. However, such flooding does not last for a long period since the surrounding water bodies are not full hence most water drains into them giving the ground enough time to infiltrate the remaining amount of water.

In summer periods, areas close to mountain, more so mountains known to contain snow might experience flooding. Areas close to glacier locations are also prone to flooding. This because as the snow and glacier melts, it needs a place for the melt water to accumulate. In most cases, the melting process occurs at a faster rate leaving the ground with minimal period for it to absorb the melt water. Such situations lead to flooding. However, this kind of flooding does not last for long periods for with time the ground will be able to absorb the melt water.

7.5. EFFECTS OF FLOODING

Flooding do affect the environment and living organisms in different ways. Some of the impacts caused by flooding are positive while others are negative. However, the negative effects far more out way the positive effects.

7.5.1. Negative Effects of Flooding

Flash flooding are known to come immediately without alerting residents. Most people are found not ready for the outcomes hence most of them lose their lives in the attempt of saving themselves from the immediate flooding. Planted crops are destroyed due to the strong forces of floodwater. Floodwater more so from ground water flooding tend to accumulate for long periods of time and this may lead to waterborne diseases some of which are quire deadly (Figure 7.8).

Figure 7.8: The immediate negative effects caused by flooding is loss of lives and destruction of property.

Source: https://www.financialexpress.com/india-news/kerala-flood-death-loss-of-property-life-and-money-rescue-operations/1285383/.

Floods do affect the livelihoods of human beings immensely. Some roads are rendered useless even after the floodwater has been drained completely. Destroyed roads caused transportation challenges affecting the sector of transport. Destruction of agricultural produce means that for the next season farmers will have to survive without harvesting any meals from their farms. Business buildings are destroyed leaving most entrepreneurs with unfit surroundings to carry out their business activities.

The general production rate of areas affected by floods drop immensely. Flooded areas are exposed to some serious vulnerabilities such as the weakening of buildings and infrastructure in such areas which does pose a great danger to the public. In serious situations, people are forces to migrate which incurs more cost for money is needed during for resettling purposes. Quite a number of the population end up dead reducing the total amount of labor in our production units hence less products being produced. Due to a smaller population, the demand for manufactured goods reduces which lowers the returns earned by businesses.

As discussed earlier, flooding lead to resettlement of people. In situations where resettlement is not a must, some people still end up moving to other locations due to loss of livelihood in their current location, decrease in production rates, deterioration of businesses among other negative impacts. A large population tend to migrate to the urban centers which are more developed leading to the rise of overpopulation issues. With overpopulation it means the general hygiene of urban centers is lowered increasing the chances of illnesses in the urban centers.

Most individuals are affected psychologically by flooding effects. Parents suffer from the loss of their children, children suffer from loss of parents, business people are traumatized by the destruction of their source of livelihood. Migration to new locations tortures individual psychologically for they have to face different challenges before they fully adapt to their new life. Loss of property traumatized many. Generally, individuals are really overwhelmed by the negative effects they have to experience as the aftermath of flooding.

The economic growth and development projects in a certain location are kept at bay for a while. Money which was initially to be used for investments is directed to helping the new homeless individuals to find a suitable shelter and basic needs. Governments end up spending huge amounts of money in the rehabilitation of flooded areas. Man power in development projects is limited for some people lose their lives as they try to save themselves from floods or as they try to sabotage the most priceless equipment from their residences. Agricultural produce to be used to earn individuals some cash flow is destroyed slowing down the growth of the agricultural sector in general. Other useful resources are destroyed by floods. Generally, floods leave a big economical gap which might take a while before it is completely filled.

The authoritative stability of regions affected by floods may arise. If leaders do not play their part in helping the affected regions to get back on their feet, residents from such areas may lose trust in them which leads to conflicts between the ones in authority and the residents. Neglecting displaced residents increases crime rates in different regions as individuals try their best to at least provide basic needs to their families. With increased crime rates means insecurity issues which causes clashes between the law and enforcement bodies and the residents. Generally, with floods, there are several political implications which are experienced.

7.5.2. Positive Effects of Flooding

Just like a coin which has two sides, flooding do also cause positive impacts to the environment, however, the positive results are quite minimal. Despite floods being quite devastating, they have always been part of nature's way of renewing natural processes which have been altered by human activities as they strive to better their livelihood. The renewal processes by floods end up having long terms effects on the environment for it takes quite a long time before humans are able to alter nature's way of life.

With flooding, several wetlands are renewed. This is because floods carry away the long accumulated floodwater in wetlands and infiltrate them with fresh water more so the groundwater type of flooding. The floodwater washed into water bodies replenished the nutrients in lakes and rivers which support aquatic plant and animal life. Natural vegetation is also boosted by floods for fresh nutrients are distributed by floodwater in different locations.

Floodwater contributes to the replenishing of soils by adding nutrients to them. As the river banks bursts, they carry along with them the fertile river sediment deposits. As the floodwater covers large tracks of land, so does the deposits from rivers get distributed. Such deposits tend to have very nutritious components which return nutrients into the soils. Floodplains along rivers, lakes, and streams which are prone to flooding tend to be quite fertile lands for during floods, the floodwater adds more nutrients into them.

As more floodwater accumulates more so with the flash flooding, the underground water is recharged and replenish. Underground aquifers contain water which feeds several streams. During the drought period, the underground water tends to reduce in volume. With flooding, fresh water sips into the ground level filling the aquifers and the entire underground water table with fresh water which ensures sufficient amounts of water are fed into streams and rivers.

Flooding activities prevent soil erosion activities and ensure that the land level is at the required position. Floodwater does carry deposits from rivers and most sloppy areas into low lands ensuring that their land level is neither too below or above the sea level. The deposits from river beds are thick and sticky enough such that they have minimal chances of being carried away as a result of soil erosion.

Flooding in the arid and semiarid areas ensures that there is adequate supply of water in such areas. The arid areas face a lot of water problems so with flooding, the underground water level is raised to a level closer to the ground leading to eruption of more oasis in deserts. The oasis provides at least some sufficient supply of water to communities residing in those areas.

Some fish species do benefit a lot from floods. The weather fish which keep on migrating from one location to another depend on floods for their survival for the floodwater tend to carry them due its high pressure and speeds to newer breeding locations. The sediments deposited in river beds by floodwater tend to be perfect breeding sites for various aquatic animals.

Some birds which feed in water logged areas benefit a lot from floods. This is because floodwater leaves nutritious fresh deposits along river banks and in water logged areas which are of great benefit to such bird species.

7.6. FLOOD PREVENTION TECHNOLOGIES

As stated earlier, the negatives effects caused by flooding tend to be more serious than the positive effects. Various technologies have been implemented to help in the reduction of flooding rates. Some of the technologies are used in safety planning for floods, others used for analysis of flood information which has been gathered from previous flooding cases, other technologies used in the general flood prevention techniques, other technologies invented for the risk management of Floods and the recovery from the previous effects of flooding (FERNQUEST, 2011).

The technologies have been developed mainly because they are not that prone to human error, they are quite accurate and they deliver results faster than humans. Just like any other manmade innovation, they do have their draw backs. Some require quite high skills for their operations. They tend to consume large amounts of power in their operations. The feeding in of wrong information into them leads to wrong results, a technique commonly referred by technicians as garbage in garbage out (GIGO)

Flood Safety Planning Technologies: It is said that prevention is better than cure. All governments do know areas which are prone to floods. It does not really require too much inventions or technologies to have safety plans in place in case of floods. Facilities which are very important to the society such as health care facilities, police stations, emergency stations, fire stations, rescue team stations among others should be located in areas which are not prone to flooding. Road structures in areas known to experience flooding should be constructed in the strongest way possible to ensure that no matter the amount of floodwater, they cannot be destroyed. Bridges should be constructed using the highest technological techniques to ensure they are not affected by floodwaters. Populations residing in areas prone to flooding need to migrate to safer lands. The floodplains could be used for other uses other than their normal uses (FERNQUEST, 2011).

Observation equipment have been invented to help in a thorough analysis of the flood heights. Such equipment analyzes flood heights of each flooding activity which does take place in floodplains. This ensures that governments are aware of the intensity of floods they should plan for.

Statistical tools have been computerized to ensure that they carry our statistics activities in a more accurate way.

Hydrological and hydraulic system have been computerized to ensure that they function at their maximum best level to give back better results to the bodies formed to deal with flood issues.

Previously, floodplains where marked on analog maps. With technology, floodplains are marked digitally in digital maps which can be accessed basically from anywhere. Any new floodplain can be detected with the digital maps making it easy for the researchers to handle flood information in the most appropriate way (FERNQUEST, 2011).

Computers are being used to help in coming up with engineering designs and the control of construction of structures that minimize or eliminate flooding activities. Robots ensure that the design from the computer is perfectly constructed to its required standards without any errors or defects.

Computerized systems have also been put into place to ensure that they alert individuals of any unexpected flooding taking place. The alarming system is designed such that it is able to detect chances of floods occurring and then alerting individuals for them to be ready.

During the planning process, governments should also consider ways in which the benefits of floods could also be appreciated. Technologies are still underway to try and help not completely eliminate flooding but find a way to manage floodwater so as to avoid it from causing any negative impacts to the surrounding. The floodwater is at least controlled with use of technological equipment to aid in recharge of underground aquifers with water, boosting of lands with nutrients, deposition of deposits which prevent soil erosion and the replenishing of aquatic life in our water bodies (stormwater.wef.org, 2016).

Analysis of Gathered Information on Floods: As stated earlier, during the planning process for safety approaches on floods, information is gathered. Such information needs to be thoroughly analyzed using the best techniques ever so as to ensure that the best decisions are made.

Technological tools have been invented to aid in the approximate determinate of the maximum water flow into a stream. Regular measurements on maximum water flow ensure help in determination of at the possible volume of floodwater which the governments have to deal with in the next flooding season. With such information, the government is able to plan the most effective way to contain floodwater so as to benefit from it and the best way to prevent it from causing negative effects to the surrounding (Craven, 2019).

Robots have been used to help in the determination of the conditions of our current drainage channels. The depth, its capacity (that is the maximum amount of water it can transport) and the speed at which the water flows in it. With this information, experts are able to determine what are the best conditions for a perfect drainage channels system that will ensure sufficient drainage of floodwater within the shortest time possible and with minimal risks or drawbacks. Perfectly computerized drainage channels are the innovated and placed in the appropriate locations. They are controlled by a machine with the help of an installed software which improves the operation speed and does its best to limit the drawbacks.

Computerized basin models have been invented to help in the determination of the amount of floodwater experienced from storms caused by the increased melting rate of glaciers and snow in the cold regions. With such information, the volume of floodwater can be estimated and appropriate measures put into place to ensure that the chances of catastrophic effects being experiences are minimal. The basin models are designed in way that they basically combine the ground surface processes in their computerized system for better results to be given back to the researchers (Craven, 2019).

Flood Prediction Technologies: As states earlier, floods, just like a coin, do have two sides which have different impacts to the environment at large. One side does have positive impacts while the other side has negative impact. Since both sides are critical in their own way, they need to be handled with a lot of care, for mismanagement of either of the sides might have very serious consequences in the end. That is why great minds have come together to try and solve this Pandora box. Innovations have been discovered to help in the prediction of floods. Technological innovated tools have been designed to their level best, and they still continue to be perfected with each passing day to ensure that that they deliver the best results possible, for with mother nature, accuracy is a thing which can barely be achieved (Craven, 2019).

With proper prediction of floods, appropriate measures are implemented. Farmers are warned in advance to move their livestock to safer locations and advised to find better grounds for their planting purposes. Critical facilities such as hospitals, rescue teams, fire fighters among others are perfectly with the appropriate materials to help in handling the negative impacts which might be caused by flooding. Incase it's a new location which might experience flooding for the first time, individuals are moved to safer grounds early in advance to prevent loss of lives.

The predictions of flooding activities do require a lot of information from various sources. Historical information on past floods such as the duration of previous floods, floodwater volume and the inundated area is of great benefit. The historical information should be combined with current data. Current data should include the volumetric capacity of streams around floodplains, the maximum flow of water the streams and the surrounding reservoirs can accommodate before causing floods, the maximum amounts of water aquifers in flood plains might occupy and the current volume of water in the underground are all important in the prediction of floods (Craven, 2019).

Weather forecasting information is also important. This is because, flood forecasting basically deals with determination of the possible total amount of water to be received from various sources and whether there is adequate space for its accumulation, and if it's not, it means that flooding activities are likely to occur. Information on the precipitation volume to be experienced is recorded, its duration for flash floods occur due to heavy rainfall within a brief period, and the weather conditions which might accompany the precipitation for most coastal floods are triggered by wind.

Flood forecasting also ensures that the most endangered areas are identified. With such information, appropriate measures are put in place. The duration it might take before floods strike a certain region is also estimated in flood prediction to ensure that residents are ready for anything that might take place. With the computerized tools, durations can be perfectly estimated, even if it means a duration of one hundred year or even more (FERNQUEST, 2011).

With flood forecasting, the technology has been advanced such that floods prediction information can be viewed from anywhere in the globe via a mobile phone with the help of the global flood monitoring system. The global flood monitoring system uses weather forecasting information stored in a central location, a satellite (a device located in space but can be used to monitor literally anything in the earth's surface). The weather forecasting data has information on precipitation which is incorporated with the vegetation data (that is determine whether a certain location has vegetation cover), soil type (cause if it can promote flooding activities), saturation level of underground water and the volume of water in surrounding streams. With all this information available in one, all one has to do is access a map via the Internet, feed in the name of their current location and the other required information, then the rest of the data is obtained from the satellites and results given back to you after thorough analysis. Use of scientific visualization is incorporated in the entire analyzation process to help in coming up with a clear picture of what might occur by combining post-data and pre-data of previous flooding activities. All this is done with only a touch of a button on your device without any need of having to do the research on your own (FERNQUEST, 2011).

7.7. METHODS OF PREVENTING FLOODS

For ages the human race has done its best in coming up with ways to try and prevent or even manage floods. Some have chosen to plant vegetation cover which reduces the speed of surface runoff and allows if to have adequate amount of time for it to infiltrate into the ground for its retention. Terraces have been constructed on the hillside to reduce the flow of water on the sloppy lands and they do also reduce the rate of soil erosion on such areas.

Manmade channels have been constructed all over to try and divert the direction of floodwater into other sides.

All the discussed above practices have been of help in one way or the other although they are not that effective. Large volumes of floodwater could not be prevented or managed using the discussed methods above. Therefore, great minds have come together so as to come up with better and unique ways to try and come up with the most effective ways of preventing floods or mitigating the aftermath of floods (stormwater.wef.org, 2016).

One of the ways used to control flooding is the construction of dams. A dam is basically a barrier like structure such as a strong wall constructed across a river in order to stop or reduce the amount of water from flowing to the other side of the river. Such a wall creates kind of a reservoir. For the purposes of controlling floods, the wall is constructed to a certain height that will prevent any water from flowing over it. In situations of heavy rains or high melt rate of glaciers and snow, the floodwater is accumulated in the reservoir for it cannot cross to the other side. Such accumulation prevents overflow of water from the river into the surrounding areas hence preventing floods. For the dams to be more effective, technological skills have been incorporated in the construction of modern dams. They are being equipped with systems that help in the collection of data such as the extra amount of water accumulated that could lead to floods, its strength, its main sources among others. The release of water to the other side of the stream is also computerized. Software's have been made to control the rate of release to the other side of the constructed reservoir. Monitoring system have also been put into place during construction of dams to aid in the management of the constructed walls. This ensures that the walls are maintained at their maximum operation levels reducing chances of collapsing. The constructed reservoirs not only serve as a means of controlling floods, but the accumulated water can be used in the production of hydroelectric power, for irrigation purposes, fish farming among other uses. However, some dams are being constructed purely for controlling floods and they are referred to as dry dams (stormwater.wef.org, 2016) (Figure 7.9).

Figure 7.9: Computerized diversion canals have been innovated to aid in the redirection of excess floodwater into temporary built ponds or water bodies with minimal chances of experiencing floods.

Source: https://civilseek.com/head-works/.

The canals are constructed in such a way that they are able to control the flow rate of surface runoff from heavy rains. They can also withstand the pressure of such waters. Canals meant to empty the water into water bodies are able to evaluate the maximum capacity level of the river, and if the capacity is quite low, they have an inbuilt system which prompts them to automatically close their one end and alert their main controlling system. Most diversion canals, have been designed with subsystem for emergency purposes. The subsystem empties the water in the main canal to the nearby constructed ponds. The ponds facilitate the treatment of such water and then it is later pumped into households to be used for domestic purposes. The water from the temporary ponds can be used for irrigation purposes. The main task is that the pond has to be empty after a certain period for its reuse. Since most of the water being redirected by canals is dirty, treatment processes are able to be conducted in the canals as the water flows to its destination. Such processes have been automated because of the soft wares incorporated in their construction.

Floodwater can be diverted into wilderness areas which can be turned into floodplains. As we all know, most floodplains are sparsely populated and in most cases, they are completely deserted. Such areas could be of great benefit in controlling floods. Diversion system are constructed to divert the any excess water in rivers and other areas affected by floods into the wilderness. Technology comes in to control the release rate of water into such areas with the help of programs incorporated in the diversion system. This is because, uncontrolled release of excess water will lead to more flooding in such areas for the rate of release will be high compared to the absorption rate into the ground. With controlled release, the ground level will have adequate time for it to absorb the water being released (stormwater.wef.org, 2016).

Floodwater can be used to automatically replenish underground aquifers with water. Technological deep infiltration system has been out into place. The excess water from rivers, glacier, and snow melt among other sources is controllably infiltrated into the underground aquifers. The infiltration system is constructed strongly to ensure that the can withstand the frequent natural changes which occur in the underground level and they go deep below the underground water level to ensure no evaporation of the water from the ground during the quite hot seasons. They are incorporated with water treatment systems for treating the excess water. This is because with flowing excess water, there is no time to collect it and treat it first. The replenishing of ground water ensures there is constant supply of water in all seasons.

Most rivers are prone to flooding. Initially, levees were constructed manually along rivers to try and contain the excess water within the river channel. However, with technology, well-engineered technologies are designed and constructed away from the river and later installed along the river banks. They are constructed in such a way that they are incorporated with system which can measure the pressure of excess water and in turn produce equal pressure to prevent them from collapsing in situations where the excess water pressure is way too high. Weirs are all being installed in rivers to try and control flooding. Weirs are basically low wall-like structure whose main purpose is either to redirect the flow of water or control its flow. They reduce the speed of excess water. Modern weirs have been incorporated with systems which can detect when the water volume is high than usual. Since most of them are movable, once the water levels are high, the weirs automatically sink themselves into the lower levels of a river to prevent flooding in the upstream. At the lower levels of the river they control the speed of the water and reduce its pressure such that it does not exceed the river channel's maximum pressure level (stormwater.wef.org, 2016).

Coastal areas have been experiencing serious cases of flooding more so with flooding types such as catastrophic floods and tsunamis being dominant in such areas (Figure 7.10).

Figure 7.10: Sea walls have been designed and constructed with the best technological expertise being used to try and reduce the flooding rates in the sea shore.

Source: https://www.reuters.com/article/us-indonesia-infrastructure-floods/in-flood-prone-jakarta-will-giant-sea-wall-plan-sink-or-swim-idUSKCN1BP0JU.

Beach nourishment systems, most of them which have been computerized, have been put into place to aid in the renovation of already damaged beaches and the detection of possible floods. Barrier islands have been created in the sea to help in controlling spring tides and storm surges all of which cause catastrophic floods. The islands have been incorporated with systems that can detect the rising of tides and the formation of storm surges due to strong winds. Once possible storms have been detected, certain tools are automatically set ready to aid in the reducing the strength of such storms and tides way before they reach the sea shore hence preventing chances of flooding. Dykes and culverts have also been put in place to help in the removal of any accumulated water due to minor floods in the sea shore areas. The dykes and culverts have been computerized in such a way that they are able to operate within the shortest time possible to try and mitigate aftermath effects of floods. Tide gates are used together with dykes and culverts to control the rate at which the water is deposited back into the sea with minimal pressure to avoid exciting the water which might lead to flooding. Tide gates can also be placed in estuaries to control the rate of release of water from rivers into the sea. This is because, if the rate is high, it means the water will be released under high pressure and may stir up the sea water and cause floods (Craven, 2019) (Figure 7.11).

Figure 7.11: Self-closing flood barrier is one of the techniques which have been implemented to help in the control of floods.

Source: http://www.floodcontrolinternational.com/PRODUCTS/FLOOD-BARRIERS/self-closing.html.

Self-closing flood barrier mostly control inland type of flooding hence they are designed and constructed in the inland areas. Their lengths do vary. Initially, they were designed to close using a closing mechanism deployed by the floodwater but since the floodwater pressure maybe too intense leading to their destruction or some passage of water, most of them have been computerized. By being computerized I mean that they are able to detect floods and automatically close themselves. Around the barriers, diversion systems have been installed to automatically divert the collected water either through pumping techniques or any other means. The diversion is mainly to ensure that the pressure of the accumulated water does not exceed the pressure of the barrier and cause collapsing effects. However, flood barriers are currently being designed with a technology that is able to immediately calculate the pressure of floodwater and in turn automatically generate equal and opposite pressure against the floodwater preventing it from causing collapsing cases.

Since floods do not last long, temporary flood barriers maybe installed in areas more so along rivers which are prone to flooding. Most temporary barriers are made of rubber materials. They operate using the techniquc of making use of the accumulating floodwater to control the floods. The rubber materials is designed in such a way that it slowly absorbs in the excess water until it is fully infiltrated and forms a levee like barrier. The strength of the rubber material is enhanced with the help of programs incorporated in its construction such that it is able to control the excess water in the river from flowing out of the river and cause flooding (Craven, 2019).

7.8. RECOVERY STRATEGIES FROM FLOODS

Not all floods around the globe could be prevented or even properly managed. The floods which manage to outdo almost all the strategies put in place to control floods do leave behind some serious negative effects. Strategies have been implemented to aid in the recovery from the negative effects caused by floods.

One of the most common way of receiving from floods is total retreat from the floodplains. Populations are supposed to migrate completely from floodplains and let them turn into wilderness. Any buildings that remained should be done away with completely and the services offered in such areas aborted. If the floods could at least be preventer in future, the floodplains could be rehabilitated into parks. However, the parks should not be filled with wild animals. The total number of animals should be controlled through transportation of the excess number into new location. The controlling of animal numbers is mainly to ensure that quick rescue missions are conducted in future in situations where the prevention strategies in place fail and cause floods (Craven, 2019; FERNQUEST, 2011).

In urban areas which are quite prone to floods, resilience could be a key strategy in recovering from floods. Resilience means that building should be constructed strongly such that they are able to overcome the strong forces of floodwaters and they still remain standing. Technology is being incorporated in the construction of resilient building to help in their perfection. With technology, robots controlled by installed programs could be used to prevent flow of water into the vulnerable areas of a building. Still, the most delicate parts of a building could be constructed in higher levels which are secure from floodwater despite the intensity of the floods. Tunnels are incorporated in the construction process of buildings to ensure that excess water is drained with immediate effect hence saving the building from any damages. Before construction of a certain building, thorough research and effective planning should be done to help to ensure that the building level a certain height above the ground level to improve the resiliency of the building recovery strategies from floods (Figure 7.12).

Figure 7.12: Alarm system for floods.

Source: https://www.deccanchronicle.com/current-affairs/280116/flood-monitoring-device-to-avert-train-mishaps.html.

Alarm system have been implemented in various parts of the world which are frequently affected by floods. However, such systems have not been that effective for in most cases they alert residence in periods which are way too late for any measures to be put in place. It is high time the alarming systems be effected such that they are able to obtain firsthand information from flood forecasting stations and alert the residents early in advance to avoid the last minute rush of trying to sabotage the critical things.

It is true that flood prevention techniques have been implemented by various governments. However, such projects are still not completely operational. Most governments have been quite reluctant when it comes to such projects. The total funds required to aid in completion and for such projects are not being given out in the appropriate amounts hence the projects are left incomplete with minimal resources and unpaid human labor. It is high time reluctant governments consider the issue of floods a serious threat and take responsibility in funding the different flood mitigation techniques which have been put into place so far (FERNQUEST, 2011).

It is high time we as humans realized that we are the major causes of most of the natural disasters occurring in our planet. We have chosen to live a stress free life. We prefer perfectly manufactured products to the manmade products. We have become way too fancy such that we ate running for the latest trends and leaving behind the old durable trends of different products. Such habits have increased the manufacturing processes in our society. Manufacturing activities lead to the release of poisonous gases which promote the depletion of the ozone layer. With the ozone layer having been depleted, the entire globe is basically turned into a small heat island. Increase in temperatures increase the melting rates of glaciers in both the south pole and north pole hence rise in the sea water levels which might cause flooding. Deforestation has increased the chances of floods due to lack of vegetation cover which retains water in the ground. It is high time as humans we retrieve from activities which are promoting climate change, for with more climate change it means more chances of floods occurring hence more hazards facing us.

Wetlands need to be protected. The practices of rehabilitating wetlands and using them for agricultural practices is one of the major reasons why floods are increasing. Wetlands need to be maintained in their perfect condition to encourage more water retention which reduces the negative effects caused by floods. If possible, trees should be planted in wetlands to improve the water retention rates (stormwater.wef.org, 2016).

With ages, the human race has done its best in increasing its adventurous missions around the world. Such practices have immensely become part of the top priorities to human race. It has led to historical straightening of river channels to improve their navigability. Manual straightening of rivers has in one way or the other affected the natural course of a river. Such changes have weakened most river systems and exposed them to various vulnerabilities. One of the vulnerabilities is lack of the required capacity to hold certain amounts of water during heavy rains or high melting rates. With low capacity means that we flood will occur. To reduce the hazardous effects of the aftermath of floods, it is recommended that individuals invest in construction of meandering structures in our rivers to improve their water holding capacity hence reduce floods (stormwater.wef.org, 2016).

Soil mismanagement practices have generally increased the chances of experiencing floods in a certain area. Mismanaged soil is very thin such that instead of absorbing moisture, all it does is release it making the water flow and accumulate in large quantities above the ground causing floods. It is high time soil needs to be manages properly by planting vegetation cover, or covering it in mulch or adding organic manure to the soil. Well managed and drained soil is able to absorb huge amounts of moisture, hold it and let it seep into the ground level preventing it from running into rivers.

7.9. FLOOD SAFETY CLEAN UP STRATEGIES

Appropriate strategies need to be implemented to ensure the safety of individuals during and after floods. As we all know, the negative effects of floods are more serious and can lead to mass death of different individuals.

The general public should be educated on the proper practices to be carried out during and after flooding activities take place.

It advisable for households in areas which are more prone to flooding or close to floodplains to always have emergency supplies ready in all times. This reduces the rate of panicking among individuals for they can be sure that their most basic needs will be met in situations where unexpected flooding do occur.

Areas which are at the risk of experiencing floods should have perfectly functioning communication system. Such systems ensure that individuals receive regular updates of what going, what to expect and any new plans which need to be put into place.

Floods do cause distraction of property. It is advisable that the most critical documents and other materials to be located in the most secure places even in areas which are not prone to flooding. Such practices ensure that important information is not lost as a result of the aftermath of flooding.

In situations where unexpected flooding does occur, it is advisable that households bring in the movable outdoor possessions into the house or place them in an encloses area. This prevents them from being carried away by floodwater.

Floodwater may lead to electric power disruptions or even destruction of gas supply systems or petroleum supply systems. During floods it is advisable that the main power supply switches be turned off, gas valves be switched off among other systems to prevent more serious disasters from taking place.

If possible, in situations where the residents where earlier informed of possible floods, it is advisable they pay attention to such warning and move to safer locations as early as possible to prevent more loss of human life which can be caused by floods.

After flooding has taken place and the floodwater is all over them place, certain safety measures must be implemented.

Driving on flooded roads should be completely avoided for a small amount of moving water might trigger the driver to lose control of the vehicle hence accidents occur or even the vehicle is carried away into nearby water bodies together with its occupants.

Floodwater is extremely dangerous. It should not be used for domestic purposes such as the cleaning of utensils, cooking, drinking or any other purposes. This is because floodwater does interact with several materials some of which might contain toxic materials. The mixture of toxic materials with floodwater end up contaminating the water making it unsafe for use. It is advisable for people to use clean treated water during floods.

After floods, most individuals who had been evacuated from flooded areas opt to return to their homes with immediate effect. Such practice is not safe unless it has been authorized by the people in authority. This is because the aftermath of floods are quite serious and such locations may take sometimes before they can be considered as safe locations for residents to reside in. After floods, thorough cleaning of our homesteads should be done. Bleach and detergents should be used in the cleaning process so as to disinfect all the surfaces for as discussed earlier, floodwater is always contaminated and it does end up contaminating the surfaces it comes into contact with. The items which cannot be cleaned should be thrown away for might cause diseases to people. In situations where the floodwater is still accumulated in large amounts, helps should be sought to aid in the restoration of our homes into perfect conditions for residing. Sometimes mold might have grown due to the accumulation of water for quite some time hence protective clothing should be put on as we try to retrieve our homes from the aftermath of floods. The protective clothing has been designed with quite high technological techniques such that they are able to detect the toxicity of aftermath areas and warn experts from exposing themselves to such areas which might be quite fatal.

7.9.1. Flood Regulations

Governments have done their best to try and prevent floods and find ways to better ways of receiving from the uncontrollable floods.

Regulations have been put into place to try and limit the human practices that are the main factors contributing to the occurrence of floods. Such practice ensures prior treatment of waste materials more so the gaseous waste materials into the atmosphere. This ensures that the ozone layer is at least preserved reducing the temperature rates on the earth's surface. Failure to observe these regulations, parties involved have to face the law.

Flood insurance policies have been implemented in most areas. The policy majorly focuses on the reduction of the insurance money paid to insurance companies and increases in the amount of government annual budget money going into flood insurance companies. With more insurance money, negative effects experienced from floods are motivated within the shortest time possible for the appropriate amount of money is available.

CHAPTER 8

Environmental Data Science

CONTENTS

8.1. INTRODUCTION

Remote sensing is referred to as the science of obtaining information concerning the earth through the use of high-flying air crafts and satellites. It involves data being collected by sensors that are attached on to the aircraft. The sensors detect the energy reflected from the earth. It can be classified into two. It can be passive or active. In the case of passive sensing, it involves response to external stimuli recording the natural energy emitted by the earth's surface. In this case, reflected sunlight is the common radiation detected by the sensors. On the other hand, active sensors make use of internal stimuli in the collection of data. This is greatly demonstrated when laser beams are used and the data collected by the projected laser on the earth's surfaces. It measures the duration taken for the beam to be reflected back to the sensor. ESR is a tool used in conducting environmental management. It allows the use of the latest technology being used in collection of data. As this is an area under study and has positive outcomes when employed, ESR has been introduced in schools (Weng, 2011).

This is a strategy that has been brought about to aid in matters environment. It is designed to facilitate the introduction of students to remote sensing science and technology. The main aim of the ERS is to place emphasis on mastering fundamental remote sensing concepts. It also looks into making adequate use of the remotely sensed data for environmental information extraction as well as problem-solving. It therefore allows the students to develop an understanding and knowledge of the principles as well as the applications of remote sensing. This includes matter-energy interactions, satellite multispectral data set, radiation transfer theory, computer-aided analysis, image interpretation, and remote sensing applications. They will also be able to provide a survey in relation to the concepts as well as techniques of remote sensing and image analysis commonly used in mapping as well as the monitoring natural resources, land use, environment, and the broad spectrum of the geoscientific applications that range in line with the meso-to global scale.

It will, therefore, go to show how remote sensing can be utilized as a tool of geo-exploration. ERS is therefore incorporated into the learning curriculum. When taught to students as a course, some of the objectives include the following:

The students being able to understand the important concepts as well as the principles of remote sensing. They should be able to know the advantages and limitations of remote sensing. They should identify the method and techniques used in remote sensing. They are required to apply remote sensing techniques to activities such as remote sensing inventory, monitoring, and analysis. The analysis in this case is geological analysis which may range from laboratory mineral rocks, ground truth, air, and space-borne remote sensing. They should be able to understand the potential of contemporary image processing as well as the analysis system. They are able to choose remote sensing data and analysis approaches on the basis of the problems to be solved. They should be able to negotiate between the interface between geospatial technologies and remote sensing. They will need to plan and also implement a remote sensing project as well as understanding the future sensing systems and application (Weng, 2011).

Generally, ERS has a broad range of applications when it comes to matters of environmental planning as well as management. Some of the areas of application include coastal application, ocean applications, hazard assessment and also natural resource management. In most of these areas, it has involved the measuring of temperature, charting wildlife habits and tracking the impact of natural disasters. In involves the conducting of various types of analysis tnat led to the production of results that will allow effective planning as well as proper management of water and land.

8.2. IMPORTANCE OF ENVIRONMENTAL REMOTE SENSING

It allows systematic data collection through the use of the latest advanced technology available. It facilitates repeatability of studies that have been conducted concerning the environment. It facilitates a global coverage more so dealing with climate and environmental conditions globally. It creates accessibility in areas that have been rendered inaccessible. In some situations, it has been proven to be the only method that can be used in conducting the studies. It also allows the obtaining of information that is multipurpose in nature (Weng, 2011).

To have a better understanding on ERS, it is important to know the concepts and principles underlying it.

8.3. PRINCIPLES OF REMOTE SENSING

Some of the principles of remote sensing include the detection and discrimination of objects or surface features through modes of detecting and recording of radiation of radiant energy that is reflected or emitted by objects present on the surface. This is illustrated by the sensors placed on the high-flying objects picking up signals or radiations from the earth and recoding the acquired data (Skidmore, 2017).

The other principle is that different objects return different amounts of energy in variety of bands of the electromagnetic spectrum that is incident upon it.

Also the radiance of the energy will vary depending on the property of the material such as its physical, chemical, and structural composition. It is also dependent on the surface roughness, angle of incidence, wavelength of radiation and its intensity. The interaction of the radiations with the object will also vary depending on the properties of both the radiation and the object (Skidmore, 2017).

The electromagnetic waves will have different characteristics in their patterns as they travel through the atmosphere. The wave will, therefore, have a characteristic shape and length. The wavelengths will also vary. Depending on the wavelength of the electromagnetic waves present in the electromagnetic spectrum the uses of the waves when it comes to remote sensing will also vary (Skidmore, 2017).

The information obtained by the sensor will depend on the distance present between the target and the platform. This implies that the distance can affect the information obtained.

8.4. CONCEPT OF REMOTE SENSING

If an electro-magnetic radiation is considered. It is either reflected or emitted from an object. The object is considered to be the source of the remote sensing data. This means that the various media available can be utilized in remote sensing. In the remote sensing, the devices used to detect the radiations are referred to as remote sensors. They can either be scanners or cameras. The transporter or vehicle used in carrying the sensor is referred to as platform. They are in most cases high flying objects such as aircraft or satellites. Remote sensing has led to various things such as photogrammetry, photointerpretation, and photo-geology among others (Weng, 2011).

The properties of the object emitting the radiation can be determined using emitted or reflected electromagnetic radiation from the object. The properties of the object tend to be unique and therefore vary from one person to another. Remote sensing is therefore the technology that has been used in detecting the type of environmental condition on the basis of the emission or the reflection. Thus the name environmental remote sensing (Skidmore, 2017).

8.5. PROCESS OF ENVIRONMENTAL REMOTE SENSING

Energy source: It is also referred to as the illumination. This is an essential component. It is the source of the energy that illuminates and provides electromagnetic energy focused on the target of interest.

Radiation and Atmosphere: This part involves the energy from the source traveling to the target and comes into contact with it. It then interacts with the atmosphere as it passes through. The interaction may take place the second time as the radiation travels back to the sensor.

Interaction with the Target: As the energy or radiation makes its way to the target passing through the atmosphere, it interacts with the target. This is, however, dependent on the properties of the radiation and the target.

Recording of the Energy by Sensor: This occurs after the energy has been scattered or emitted by the target. The energy travels back to the sensor which is remote and is not in contact with the target. The sensor then collects and records the energy from the target.

Transmission, Reception, and Processing: Once the energy has been recorded by the sensor, the sensor has to pick up the duty of transmitting the recorded energy. The energy is transmitted in waveform to a receiving and processing station. Here, the data is processed into an image either in hardcopy or digital copy.

Interpretation and Analysis: This step involves the obtained data being interpreted. The image obtained is interpreted either electronically or visually. This is with the aim of extracting information concerning the target that was illuminated.

Application: This is the final element of the process of remote sensing. The obtained information is applied to the extract from the imagery concerning the target. This facilitates a better understanding of the information and to some extent, some information will be revealed allowing the solving of a given environmental problem.

This is an outline of the stages followed in remote processing. It is the actual sequential order that is followed from beginning to the end.

Below is an outline of how the ERS has been used in the different areas in the environment. It includes geographers being able to identify changes on the earth's surface that may require to be mapped. In this case they may make use of photographical data that has been obtained. It may also involve the use of satellites. The foresters are able to obtain information on the type of tree to be grow or those that are growing. They are also able to identify whether they have been affected by diseases or fires. The environmentalists when they want to detect or identify the movements of the pollutants such as oil slicks on the oceans. Geologists are interested in mining, are able to identify valuable minerals. This avoids wastage of resources and also destruction of land when identifying the areas having the minerals (Weng, 2011).

For farmers they will be able to keep watch of their crops as they grow and is able to identify when they are affected by drought, floods pests or diseases. For the ship captains they are able to identify the best route while in the northern ice packs. Firefighters are able to send out their crews on the basis of the information that they are given concerning the size and movement of forest fire. For the urban planners they are able to map and monitor the land cover, morphology, land use and the process of their formation and transformation among others (Weng, 2011).

Under ERS, there is data recording. Data recoding in most cases makes use photographs. An image is a two-dimensional representation of objects in a real scene. The remote sensing images are a representation of the parts of the earth surface depending on the selected point. In the case of a digital images, they compromise of two-dimensional array of individual picture components referred to pixels usually arrayed in columns and rows. Each pixel gives a representation of a certain area on the earth's surface. A pixel is composed of both an intensity value as well as a location address within the two-dimensional image (Weng, 2011).

8.6. USE OF ERS IN ENVIRONMENTAL RESOURCE MAPPING AND MODELING

The information obtained from ERS has been greatly used in urban studies, hydrological modeling, fractional vegetation cover and impervious surface are mapping. In matters of environmental studies, the remotely-sensed data has greatly been used in natural resources mapping as well as the source of input data for environmental process modeling. The remotely sensed data can be sourced from a variety of platforms within a broad range of spatiotemporal, spectral, and radiometric resolutions. This data has been very useful more so when used in large scale studies and applications. There are a variety of applications. They include hydrological modeling, energy, and water flux estimation, watershed mapping, fractional vegetation cover, urban modeling, impervious surface modeling, and drought predictions based on the soil water index. There are also some developments being done on the sensors so that the quality information is improved (Chang, 2012).

There are usually eight eras of remote sensing. There are some that run in parallel period times. For these ones, they tend to be unique to the terms of the technology in use, application in science, and the data characteristics. The distinct eras are outlined below:

Air Borne Remote Sensing: This era was utilized during the period of the Second World War. During this time, remote sensing was used for the aim of conducting surveys, reconnaissance, military surveillance and mapping.

Rudimentary Space Borne Satellite Remote Sensing: This era was as a result of a test concept on rudimentary satellites. This was followed by the first meteorological satellite commonly referred to as Television and also the Observational Satellite.

Spy Satellite Remote Sensing: At the peak of the cold war, spy satellites were greatly utilized. This allowed the collection of data that was mostly used for military purposes. The data use in this era was not digital. It was produced in hard copies. Over the three eras, the remote sensing shifted gears. This implies that instead of it being used for military purposes, it was being used in environmental and natural resource applications.

Meteorological Satellite Sensor Remote Sensing: The meteorological satellite sensors used were used in environmental satellites. This era involved data being availed in digital format. The data was also analyzed using exclusive computer hardware and software. This era so to it that global coverage was achieved as it was realized. This allowed environmental applications to be more practical.

Landsat Era: This era started with the launch of the Land sat 1 that contained a multi-spectral scanner sensor. Other satellites were also achieved. However, the Land sat 6 failed during its launch. The next one (Land sat 7) was an improvement of the initial one. The Land sat is the satellites that are in charge of gathering data on a global scale. This satellite has allowed a larger range of environmental applications both locally and globally. It is therefore a significant era.

Earth Observing System Era: This satellite started with the Terra satellite. It has achieved certain milestones in matters global coverage such as frequent repetition of coverage, high-level processing, and also data has been made to be freely accessible by various users. The satellites carry sensors such as a moderate resolution imaging spectroradiometer among others. The sensors used have a variety of applications that are more compared to the initial applications. It has also allowed the data to be freely accessible to all individuals and currently it is said to have over 40 products.

There is also the active spaceborne remote sensing that utilizes radar technology. This is among the latest trends and is currently being used by most of the individuals. There is the Shuttle Radar Technology Mission that is currently being used on gathering data for digital evaluation.

New Millennium Era: These satellites employ the use of highly advanced test of concepts into the orbit. The concepts and ideas used in this satellite are generally different from the EOS era. The satellites used in this case are said to be of the next generation. It involves the use of the concept of the Advanced Land Imager in place of the Land sat idea. This is because it is much cheaper and by far a better technology to be used.

Private Industry Era: It involves all the latest innovations that have been made. It involves the collection of data in a relatively high resolution. This is well demonstrated by the Quickbrid satellite. It also involves the revolutionary means of data collection. This is demonstrated by the Rapid eye satellite composed of five satellites. They have the ability to conduct a daily coverage of the earth at any spot. There is the introduction of microsatellites which is mostly used during disaster monitoring constellation. It is the property of the surrey satellite technology Ltd. The innovation of Google is also under the technology. It is used in making rapid data access of VHRI in any part of the world. This is good possible through the use of streaming technology therefore allowing even a non-specialist to zoom and pan remote sensing data.

8.7. ENVIRONMENTAL MODELING

This area has greatly made use of state of the art satellite sensors. They have been extensively used in environmental applications and natural resources management. The produced data are usually in a wide range of scales, band numbers, radiometry, and bandwidths. This facilitates some advantages such as consistency of data, Global reach, cost per unit, precision, repeatability, accuracy, and synoptic coverage. The other advantage is that the data is freely accessible online (Chang, 2012).

In matters environmental monitoring, there is need for frequent coverage of the same area. This can greatly be achieved through the use of data obtained from multiple sensors. Before being used, the various resolutions such as the bandwidth and the conditions should be harmonized and also synthesized before they are used. This will contribute to the normalizing of sensor characteristics such as the pixel sizes, spectral domain, radiometry, and times of acquisition. There is also the use of inter-sensor relationships that will assist in the establishing seamless monitoring of various phenomenon across the landscape (Skidmore, 2017).

Impervious Surfaces: These are anthropogenic feature through which water is not able to penetrate into the soil. They include roads, sidewalks, driveways, and parking lots among others. In recent cases, it is a major indicator of the level of urbanization as well as an indicator of the level of environmental quality. There are a variety of digital sensing strategies that have been developed to be used in the estimation and mapping of the impervious surfaces such as image classification, multiple regression, artificial neural network and regression tree algorithm. There are a variety of methodologies that are employed. It also involves the use of up-to-date techniques in impervious surface estimation and mapping. This may involve the analysis of case studies as well as the examination of the methodologies that are used. The information is useful to industries, the government and also beyond.

8.8. USE OF THE ERS IN ENVIRONMENT AND DEVELOPMENT

ERS can be used in the management of the natural environment more so dealing with the biotic and abiotic environment. In the biotic environment remote sensing methodologies have been used in furnishing spatial, thematic, and temporal information concerning living organisms. The information is greatly utilized in a broad spectrum. It can be used in the detection of animals such as the large mammals and the occurrence of larger features habitat types, landscape patterns and land covers. One can also gain knowledge concerning the vegetation species, spatial patterns and the modifications that have been made in the phonological cycles. The information can be obtained from multiple vegetation indices and the ecological landscaping metrics. This is all dependent on the remote sensing inputs as well as digital image analysis (Barrett, 2013).

In the case of spatial patterning quantification in landscapes. It has tended to focus more on forestry which has been a major area of concern in ecological, forestry, and also management-based research. Among the applications dealing with generic use of the management of biological diversity consist of land cover, land use change, oil-spill cleanup, grassland conditions and wildfire recovery among others. Remote sensing has been useful in the development of a foundation for a transdisciplinary approach to wildlife management and also biodiversity. The approach is dependent on the capability to derive multispectral views of the environment in several spatial and temporal scales. These have been integrated along with other forms of data. They include the global positioning systems as well as the geographical information system. This is done with an aim of producing sustainable models and visualizations. Ultimately there is the need to explore solutions to the initially intractable environmental science and management problems.

In the aquatic environment is one that has greatly utilized remote sense applications. This is used in environmental management in matters water. The remote sensing is useful in providing information about the quality, quantity, geographical distribution and the extent. There are various ways in which the water parameters are collected. They include the situ point measurements. This method does not allow for statically significant distribution known as geo-visualizations. This is attributed to the fact that the point measures tend to be sparse and also, there is an insufficient infer regional geographic patterns. In hydraulic parameters, the important variables mostly relate to the surface extent of the water bodies, water depth, water components, water-surface temperature, cloud cover, precipitation, and water vapor. There is a challenge of obtaining all these variables through the use of situ point observations (Barrett, 2013).

Through the use of remote sensing, it is possible for one to map a river channel morphology and in-stream habitat. This is done to aid in the spatial-temporal patterns of snow cover in large areas more so in inaccessible terrain. It therefore allows the provision of useful information on a critical component of the hydrological cycle. It also allows the detection of changes in wetlands through the use time series vegetation indices that are obtained from multi-temporal satellite imagery. It also useful in monitoring the ocean's dynamic process. This includes the study of the changes of the biological, chemical, and the physical characteristics. It monitors the ocean's interactions with the atmosphere, cryosphere, and the land. All this made possible through the use of remote sensing methods.

In matters rocks and soils, it involves the geological aspect of the environment and its properties. The methods availed in remote sensing is useful in the chemical composition of rocks and minerals present on the Earth's surface that are not covered by dense vegetation. There is the use of geological remote sensing approaches and satellite imagery selection. There are also methods that allow enhanced detection of lithological as well as hydrothermal alteration patters and the regional mineral exploration. Some of the methods used have 3D features of the earth's surface that is as a result of the natural processes. Through remote sensing, processes such as geomorphology and landforms can be extracted, classified, and mapped. This may however need the use of GIS techniques (Barrett, 2013).

There is the utilization of images to assist in the measuring of the soil particles. Soil mapping involves the landform mapping. They also utilize the importance of the theoretical value of the registration of the information. The practical bearings are able to affect other entities than the non-living ones. Some individuals have tried to study asses the contribution of the geo-morphological mapping as a result of the integration of remote sensing. These approaches have facilitated the enhancing of natural resource management as well as the ameliorating land use planning in the developing countries. There are some applications of remote sensing that places emphasis on observing and monitoring built environment.

8.9. TECHNOLOGIES USED IN REMOTE SENSING

Airborne laser scanning. This technology involves the use of active remote sensing. It makes use of 3D point cloud of objects through the use of LiDAR instrumentation. The platform used in this case is an aircraft. There is the Digital Elevation Model which makes use of both the DTMs and DSMs.

There is the Dynamic Global Vegetation Model. This technology is used in the monitoring of shifts in global vegetation and associated process. The Digital Surface Model. This technology studies the elevation of the land surface and also the surface features including vegetation and buildings. The Digital Terrain Model is used in the elevation of a land surface more so the one above the sea level (Chang, 2012).

The Canopy Height Model is a method which involves the use of canopy height that was calculated from the difference between a DTM and DSM. Growing Degree Days above degrees which involves the essential amount of time present for most of the plants to be able to grow. Geographic Information System which is a system for managing and also presenting geographic or spatial data. Global Navigation Satellite System which utilizes satellite navigation systems to provide autonomous geospatial positioning through the use of global coverage. There is the Global Positioning System which is a type of GNSS that makes use of the US constellations of satellites (Weng, 2011).

Individual-based models are simulation models that describe autonomous individual organisms. Leaf Area Index is a method that use a one-sided green leaf area per unit ground surface area. There is the Light Detection and Ranging. This method involves the utilization of pulsed laser in measuring the variable distances to backscattering objects including vegetation elements as well as the bare back. There is also the Photo-synthetically Active Radiation which involves the determination of light available for photosynthesis usually within the wavelength of between 400–700 nm in the range of the wavelength.

The Species Distribution Model involves the species occurrence as well as the abundance in response to the environmental factors. The Time Domain Reflectometry which is a method that involves the use of a measurement technique used in determining the characteristics of the electrical lines through the observation of reflected waveforms. Terrestrial Laser Scanning which is an active remote sensing method that acquires 3D point clouds of objects through the use of LiDAR instrumentation. Finally, there is the Unmanned Arial Vehicles which is a methodology that makes use of an aircraft without a pilot in the aircraft. They are therefore referred to as drones (Chang, 2012).

Leaf Area Index: This is one of the approaches used in remote sensing. It is unique in that it does not make use of ground-based methods. It deciphers the reflected transmission. The plants contain chlorophyll in its communities. The pigments absorb part of the solar radiation present in the spectral bands. It utilizes two widely used infrared and red ones. It is therefore changed proportionally to the amount of green vegetation present in the plant. It utilizes the use of large-area maps of LAI which is required for global land-surface modeling. There is also the use of satellite or airborne image reflectance and the ground based estimations of the LAI. It has contributed to the generation of vegetation indices have been generated from radiances within a broad range of channels that correspond to the spectral bands.

When using the LAI in estimation involves the use of satellite data. It will require the use of ground data for validation and testing of bias. The satellite data should be corrected due to the atmospheric effects. It will therefore require the use of additional information on the state of the atmosphere. This is in reference to water vapor, aerosols, and ozone. In this case a non-destructive measurement is the most appropriate approach for obtaining the ground measurements. There is a range of indices that has been derived from remote sensed vegetation containing classical values of the LAI. There is also the fitting empirical exponential function with a plateau that indicated a saturated signal implying higher LAI. The relationship was established through the use of a study on maize and wheat while at their various stages of growth. The limitation of this method is that it only attains the level 15. The values tend to be above five.

Remote Sensing of Oil Slicks in Deepwater Horizon Damage Assessment: In matters environment, there is a need to conduct damage assessments so that they appropriate recovery tactics are employed in aiding the environment to heal from the damage. It has entailed the decision to integrate results from multiple remote sensing methods. The methods, in this case, are specific to the detection of the presence of oil on the surface. The different methods used show the significant improvements that have been made in the-state-of art though it has a high degree of uncertainty. The challenge that has been encountered is when determining the thickness of a floating oil layer.

It has also been acknowledged that there is a shortage of the qualitative assessment of the DWH surface oil constrains in matters precision of the thickness estimation obtained from remote sensing. The thickness classes are used in reflecting of the uncertainty. The priority when using the multiple available sensors on the basis of the primary on the resolution of the sensor data as well as the ability of the sensor being able to make distinction between the different oil-thickness classes. The priority for thick oil is the TM though there are no seen sheens (Weng, 2011).

SAR is the greatest priority for the total oil coverage but the least for the thick oil. This attributed to the fact that OEDA is able to distinguish particularly thick emulsified oil. In case of a large oil spill it is recommended to pick up an approach that involves the incorporation of independent data from multiple sensors. It also involves the surface oil be accessed through the use of weight-of-evidence approach. It has advised that comprehensive synoptic sampling which involves data collection from both vessels and aerial imagery. This will facilitate the provision of the field verification of the satellite imagery analyses.

The methods open up new ventures as they break up new grounds in the application of remote sensing more so when dealing with large scale oil spills. The limitation was that it was thought that remote sensing had limited capabilities when used in the location of actionable oil and also tracking its transit as it moves through the open ocean. Useful thickness information can be derived and there are suggestion of applications for the new generation of SAR sensors which are recently becoming more available. There are geospatial approaches to data processing. There is also an exemplified in the integration of multiple diverse datasets within the ERMA. There are improvement efforts that have been made in matters response efforts and post DWH research (Skidmore, 2017).

There is vast information that has been publicly archived by the use of NOAA data portals including ERMA. This will greatly benefit marine research in the different fields. There are developments of new methodologies as well as the validation that will facilitate a better approach in case of a large oil spill occurring in an area. The preparedness will be adequate in managing the situation. There are advances that have been used in environmental remote sensing. They include the use of sensors, algorithms, and other applications. They all include the use of the development in concepts, methods, techniques as well as applications that have been used in remote sensing.

8.10. RELATIONSHIP BETWEEN GIS AND ERS

There are four distinctions under the advances in in environmental remote sensing. Other than that, there is the use of GIS in handling the environment. GIS is the short form of Geographic Information System. This is a system that is computer-based which conducts the collection, analysis, stores, and also the dissemination of geographic information. This is made through the use of different entities to give a platform for an informed and smooth decision making. It is greatly used the various fields in the environment. Both ERS and GIS are both used in handling matters concerning the environment. It can be used in managing issues regarding the environment. To a greater extent, it has been used environmental impact analysis (Skidmore, 2017).

They both aid in obtaining, collection, and recoding of data. The obtained data is used in the analysis process. The various factors in the environment can be analyzed. It also allows the analysis of areas of the environment that cannot be access. With the systems used in GIS, it allows a comprehensive analysis to be conducted therefore the results that will be obtained may be more comprehensive therefore allowing the identification of other issues that may be present in the environment. In the long run, it can be used in disaster management as it provides comprehensive analysis of issues therefore allowing the appropriate strategies to be employed when handling the issues faced when managing the situations. The GIS systems as well as the ERS have been integrated and used in handling environmental disasters.

The information obtained is also used in risk management as well as the analysis identifying areas that are more likely to be prone to environmental disasters. After the identification, there are preventive measures that have been developed. It also assists in the zoning of landslides hazards. Landslide hazard zoning involves the steps used in ranking of the different parts of areas on the basis of the degrees of the potential or actual hazard from the landslides. They are both useful in the evaluation of a landslide hazard making it a complete task. They both facilitate the efficient collection, manipulation, and integration of the spatial data. They include geological, structural, slope characteristics and the surface cover of the area which is relatively important for hazard zonation (Barrett, 2013).

There is also the determination of land cover and the land use. Both ERS and GIS employ the use of technologies that may be employed in land use and land cover application allowing its determination in a variety of areas. It has been advanced to the extent that it can allow the detection and estimation of the changes in the land use and land cover pattern within a given period of time. This implies that any sudden changes in land use and land cover such as natural forces as well as other activities such as deforestation. They both are also used in the estimation of flood damage.

8.11. CHARACTERISTICS OF DATA USED IN ERS

There are a variety of characteristics of the data used in ERS. Its quality involves its spatial, radiometric, spectral, and temporal resolutions. In the case of the spatial resolution, it involves the size of the pixels recorded in a raster image. Basically, the pixels correspond to the square areas and range in side length. In spectral resolution, in entails the wavelengths of the frequencies. They have different frequency bands recorded. They are usually related to the number of frequency bands that have been recorded by the platform. In the latest Land sat collection, it contains seven bands. They include infrared spectrum having a variety of spectral resolution. There is the Hyperion sensor on the earth observing that involves the resolving of 229 bands as well as the spectral resolution (Barrett, 2013).

In the radiometric resolution the number of different intensities of the radiation. These radiations can be distinguished by the sensor. They basically range from 8 to 14 bits. They also correspond to the levels of gray scale and to the shades of color in individual bands. It is dependent on the noise of the instrument. Temporal resolution. Here, the frequencies of the flyovers by the satellite is relevant to the in-time series studies. It may also be relevant to the mosaic image demonstrated in deforestation monitoring. This was at first utilized by the intelligence community which involved the repeated coverage revealing challenges in infrastructure, the deployment of units or also the modification of the equipment (Weng, 2011).

The cloud cover over a given area or object necessitates the need to repeat the collection of the specified location.

8.12. SOFTWARE USED IN ERS

The data collected and recorded in ERS is usually processed and analyzed using computer software and are referred to as remote sensing application. There are a variety of proprietaries and open source applications in existence that promote the remote sensing of data. Some of the remote sensing software packages are as follows:

ERDAS IMAGINE which is from Hexagon Geospatial, e-Cognition from Trimble, IDRISI from Clark Labs, PCI Geomatics, ENVI from Harris Geospatial Solutions, TNT-MIPS from Micro images and Remote View by Overwatch Textron Systems and the Dragon\IPS which is among the oldest remote sensing packages still in existence. The open sources in remote sensing include Optics, Orfeo toolbox, Sentinel Application Platform by the European Space Agency, GRASS, GIS, QGIS, and ILWIS. There other development being made so that there is more software available to be used in remote sensing.

As remote sensing has been introduced in education systems, the students are actively involved in coming up with new software that works on the challenges provided by the existing software ensuring that there is a better software available for use. This will also ensure that the processing and analysis of data will be more comprehensive giving off quality data.

8.13. DATA PROCESSING

Data processing is an important stage in ERS as after it is conducted, it paves way for data analysis. It is therefore the expectation of the remote sensing systems to extrapolate the sensor data focusing on a reference point. It involves the distance between two points on the ground and it is dependent on the type of sensor used. In the case of conventional photographs, the distances tend to be accurate while at the center of the image. There is a distortion of the measurements that increases as there is a movement from the center of the image (Weng, 2011).

Another factor put into consideration is the paten against the film is pressed. This may cause severe errors in the case where photographs are used in measuring ground distances. The step undertaken is the solving the problem is referred to as geo-referencing. It involves the computer aided matching of points in the image. This is extrapolated through the use of an established benchmark which warps the image therefore producing accurate spatial data. In most cases, the satellites images are sold fully to geo-referencing. There is also the requirement of images to be radiometrically and atmospherically corrected (Weng, 2011).

Radiometric Correction: This facilitates the avoidance of radiometric errors as well as the distortions. In most cases, the illumination of objects on the earth surface is usually uneven. This is attributed to the difference in the properties of the relief. This factor is taken into consideration in the method of radiometric distortion correction. The radiometric correction gives a scale to the pixel values for instance the monochromatic scale is converted to an actual radiance values.

Topographic Correction: It is also known as the terrain correction. There is a variation in the effective illumination of pixels which is attributed to the terrain in rugged mountains. In remote sensing images the pixels present on the shady slope are known to receive weak illumination and also has a low radiance value. This is different from the pixel in a sunny slope which receives strong illumination and also has a relatively high radiance value. If the same object is taken into consideration, the pixel radiance value on the shady slope will have a difference from the one on the sunny slope. Different objects may have the same radiance values. It may be due to the fact that they have been affected by the remote sensing image information extraction accuracy in the mountain areas. It therefore becomes the greatest obstacle to further application of remote sensing images.

The reason for conducting a topographic correction is to facilitate the elimination of the effect therefore recovering the true reflectivity or the radiance of the objects in the horizontal conditions. It therefore the premise of the quantitative remote sensing application.

Atmospheric Correction: This involves the elimination of the atmospheric haze by conducting a rescaling of individual frequency bands therefore making the minimum value to correspond to the pixel value which is zero. The digitalizing of data has facilitated the manipulation of data through the changing of the grayscale values (Skidmore, 2017).

An important factor in making sense of data is its interpretation. The initial application included aerial photographic collection. This utilizes a number of processes which include spatial measurement by utilizing light table in both conventional single and stereographic coverage. Added skills which involve the use of photogrammetry, use of photomosaic and also repeat coverage. The recent developments include the use of image analysis in computer-aided applications. It has become more rampant in use. There is also the utilization of old data from remote sensing which is most valuable as it provides long term data used in large scale geography.

The old data is in most cases complex when being interpreted and also bulky when it comes to storage. The modern systems as they are able to store the data in a digital format and with lossless compression. The limitation of using this approach is that the data is fragile and the format used may be archaic. The data may also be easy to falsify. The best systems used when it comes to archiving of data series involves the use of computer-generated machine-readable ultra-fiche. It utilizes type fonts including OCR-B and also digitalized halftone images (Weng, 2011).

The ultra-fiche does well in standard libraries. They can be created, copied, filed, and retrieved through the use of automated methods. The information is about compact as the archival magnetic media. The advantage is that they can be read by human beings with minimal standardized equipment. Generally, remote sensing works on the principle of the inverse problem. As the object of interest may not be directly measured, there is an existence of other variables that can be detected and measured. This may be related to the object of interest through calculation. The analogy is commonly given to describe the principle. But this is like determining the type of animal from its footprint.

For instance, it is impossible to directly measure the temperatures of the upper atmosphere, it is however possible to measure the spectral emission from a known chemical species within that region. The frequency of the emission can be related through the thermodynamics to the temperature within the specified region (Skidmore, 2017).

8.13.1. Data processing Levels

The levels were established to aid in the discussion of the data processing practice. The levels are outlines as shown below:

The O level: This involves reconstructed, unprocessed instrument, and payload data at optimum resolution. This is along with all and any communications artifacts removed. The communication artifacts include synchronization frames, duplicated data, and communication headers.

The 1a level: This housed the reconstructed, unprocessed instrument data at full resolution time-referenced with ancillary information such as radiometric and geometric calibration coefficients as well as geo-referencing parameters. They are all computed and appended but are not applied to the Level 0 data. In the case where it is applied, it is applied in a manner in that the level 0 is fully recoverable from level 1a data.

Level 1b: This level involves the level 1a data that has been processed into sensor units such as radar backscatter cross-section and brightness temperatures among others. It is important to know that not all instruments have level 1b data. Also, level 0 data is not recoverable from level 1b data.

Level 2: Here, we deal with the derived geophysical variables such as ocean wave height, ice concentration and soil moisture at the same resolution and at a location as level 1 source data.

Level 3: The variables are mapped on uniform space-time grid scales. They usually have some completeness and consistency. This may involve missing points being interpolated and the complete regions being mosaicked together from multiple orbits.

Level 4: This level deals with model output from the analysis of lower data. For instance, the variables that were not measured by the instruments but instead are the derived from such measurements.

In summary, a level 1 data record is the highest reversible level data record. It has a significant scientific utility. This creates the foundation in which all other subsequent data sets are produced. Level 2 is the first level directly usable for most of the scientific applications. Its value tends to be relatively higher than the lower levels. The data sets in level 2 tend to be less voluminous compared to level 1. This is attributed to the fact that they are temporarily, spatially or spectrally reduced. Level 3 data sets are smaller than the lower level data sets. Therefore, they can be handled without incurring a great deal of data handling overhead.

The data in this case tend to be more useful for most of the applications. To make level 3 data sets can be made to be more feasible so that they readily combine data from various sources. This is achieved through its regular and temporal organization. These are the common levels defined in data processing. Other data level vocabularies have been defined. They may be suitable for more heterogeneous workflows (Chang, 2012).

Remote sensing has been incorporated into the daily living. This is seen by it being used in weather forecasts in the reporting of climate change or natural disasters. This is has led to it having a relevant influence in the modern information society. This is an important technology used in aerospace industry. It also bears a great economic relevance. There is a development in the technologies being used. For instance, it is used in Terra SAR-X and Rapid Eye technologies. There is studies and education being given on matters ERS. This is important as there is a big gap present between the application and understanding of the satellite images. There is computer software being developed to facilitate school lessons and in the curriculum. When teaching remote sensing in schools, it has involved it being taught through the consolidation of subjects such as math and physics. It has also involved the use of media fields taught in schools. This will facilitate a better interpretation on the satellite images. Remote sensing has been incorporated to some of the curriculums (Skidmore, 2017).

8.14. APPLICATION OF REMOTE SENSING IN THE ENVIRONMENT

The conventional radar is utilized in air traffic control, early warning and in meteorological data in a large scale perspective. There is also the Doppler radar which is utilized by the local law enforcements in the monitoring of the speed limits. It also allows an enhanced meteorological collection including wind speeds as well as the directions present in the weather systems along with precipitation intensity and location. There is also the use of lasers as well as radar altimeters on satellites in the provision of a variety of data. It facilitates the measuring of water as a result of gravity. It involves the mapping of features on the sea floor to the resolution of a mile or beyond (Figure 8.1).

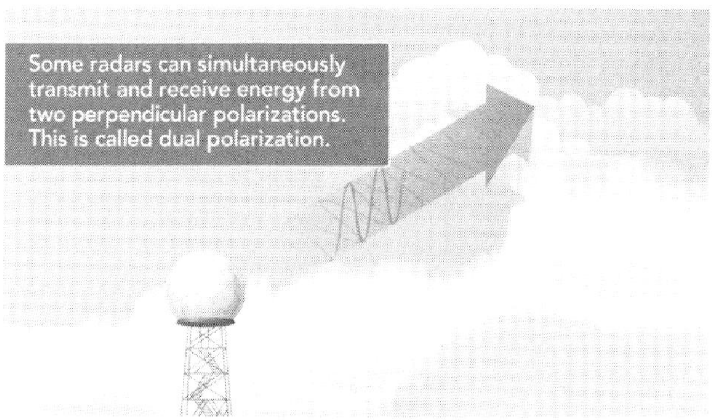

Figure 8.1: The conventional radar.

Source: https://www.youtube.com/watch?v=M7FVIbFLxfs.

There is also the measuring of the wavelength and height of ocean waves, the altimeters facilitate the measure of wind speeds as well as direction. It measures the surface ocean currents and directions. The acoustic as well as the radar tide gauges are used in the measuring of sea levels, tides, and also wave direction in both coastal and off shore tide gauges. There is the light detection and ranging which is well known for weapon ranging as well as laser illuminated homing of projectiles. The LIDAR is utilized in the detection and the measurement of concentrations of a variety of chemicals present in the atmosphere. The airborne LIDAR is used in the measurement of heights of objects and features present on the ground which is more accurate compared to radar technology. A major principal of the application of LIDAR is the vegetation remote sensing (Figure 8.2).

Figure 8.2: Applications of LIDAR.

Source: https://www.elprocus.com/lidar-light-detection-and-ranging-working-application/.

Both radiometers and photometers are the common instruments used in the collection of reflected as well as the emitted radiation in a broad range of frequencies. The most common frequencies are those of the visible and infrared sensors. They are followed by microwave, gamma ray and also the ultraviolet (UV). They may be utilized to facilitate the detection of emission spectra of a variety of chemicals therefore providing data on the chemical concentrations present in the atmosphere. The radiometers are utilized at night. This is attribute to the fact that artificial light emissions are an indication of human activity. They are used in remote sensing of the population, GDP, disasters, and damages incurred by the infrastructures due to wars. All this information is important in the management of the environment (Figure 8.3).

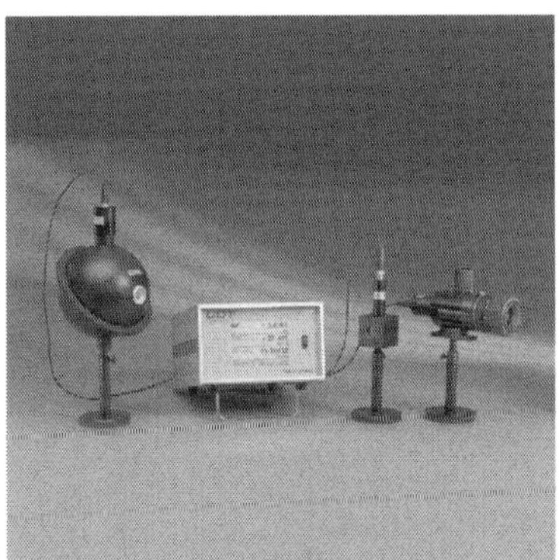

Figure 8.3: Radiometers and photometers.

Source: https://www.elliotscientific.com/Gamma-Scientific-Photometers-Radiometers-and-Photodetectors.

Spectropolarimetric imaging has been said to be useful in target tracking purposes. Man-made items are said to possess parametric signatures that are not present in natural objects. The stereographic pairs of the aerial photographs have been utilized in the making of topographic maps. This is made possible by the terrain and imagery analysts in various departments (trafficability) in identification of possible routes. They also conduct terrestrial modeling of habitat features. There is also the Hyper spectral imaging which produces images where the pixels contain complete spectral information. The imaging allows the narrowing of the spectral bands over a contagious spectral range. It is utilized in mineralogy, biology, defense, and the environmental measurements.

Apart from the meteorological used of remote sensing in the environment, it is also useful in agriculture, biodiversity, and forestry among others. For instance, mining is a major part of the environment. If the mining sites are well identified, it can allow the protection of the environment. Remote sensing has been used in 3D mapping for uranium enrichment. There are multiple satellites used in the acquisition. There are also 3D mapping systems which are useful in differentiating between the volumes. This technique has proven to be successful and is being used in several states.

The roads are also part of the environment. The roads need to be assessed so that the appropriate measures are taken to ensure that they are well maintained. The various remote sensing techniques are being used in the assessment of the roads. They also make use of GIS techniques, therefore, acquiring high levels of accuracy. This allows several resources to be saved in terms of time and money. The fuel production affects the level of certain chemical composition in the atmosphere. The satellites have been equipped with the capabilities of measuring the vehicle emissions such as CO, HC, and NO among others. This technique has been utilized by the government by allowing them to put pressure on vehicle owners to follow the emission standards.

There is need to protect the landscape from deformation. Remote sensing has been used in the assessment of train stability. There is the utilization of the interferometer technology which allows the measurement of deformation through the use of different phases. This method is used in the sectors in the gas and oil sectors. When dealing with snow, there is the calculation of snowpack. This allows researchers to understand the ratio in which snow melts. This involves the utilization of LIDAR technology together with a spectrometer to facilitate the absorption of sunlight. This facilitates a much better understanding.

When looking at the state of the earth, images may be useful in showing the nature and state of the environment. Remote sensing has been useful as it has allowed the collection of the earth's pictures from space. This has utilized satellites to produce the images. The satellites transmit the images to computers which allows the production of a hard or digital copies of the images. The use of digital copies is better as it allows zooming in on the pictures. It is also useful in the conservation of lakes and rivers.

The conservation of lakes and rivers involves the wooded areas along the waters, therefore, protecting the lakes and rivers from running off. The remote sensors allow the identification of the riparian zones which may need assistance. This may involve the use of high-resolution satellites in the observation as they spread across the globe. Controlling of forest fires. This has been made possible through the use of data that has been acquired through the use of remote sensing. The firefighters have been given this information and have been able to control the fires in the specified locations.

In matters of animals, it is important to know the numbers of the animals so that one can identify those that are endangered. For instance, remote sensing has been used in the counting of polar bears among other animals that have been considered to be in extinction. The ecologists make use of the information obtained from the sites. The advantage is that the data obtained can be trusted therefore research can be conducted and the appropriate measures to be taken therefore ensuring that the animals do not go to extinction.

There is the estimation of the forest supplies. This involves the use of AVHRR, SPOT, and MODIS have been regularly utilized to measure the increase in the global forests as the forests are known to be a major source of valuable materials. This may be attributed to the fact that the satellites are useful in carrying out a surface evaluation. The evaluation is done using a technology called the inter-aerometric synthetic aperture radar.

A part of the environment is the arctic region. It has been known to greatly suffer from global warming as most of the ice has melted. The satellites have been used in the exploring and navigation of the artic. The remote sensing technologies conduct activities such as ice monitoring and ship tracking giving off data concerning the arctic region. It also conducts an evaluation on the environment identifying whether it is suitable for animals. For instance, the level of pollution has greatly affected aquatic life.

Once the analysis is done, appropriate measures can be taken to protect the environment and improving the state of the environment. This will allow the animals to survive in the given environment.

It has been known that some species of fishes are heading to extinction. This includes shark among others. Biologists have made the use of satellites to track the movement of these animals. For instance, in the case of sharks, the biologist may attach a satellite tracker on fish. They then track the movements of the fish. They then obtain data concerning the fish by so doing they are able protect the species of the fish. The remote sensing technology has been used in conducting observations on the surface color as well as the temperatures of the given area.

In the long run, they are useful in the preservation of marine life as well as the environment. This may be attained through the prevention of oil slicks which may be used in preventing damage to the marine life as well as the surrounding. The remote sensing technologies can be used in the location of oil spills as well as the determination of oil direction as well as the spread ratio.

In summary, the use of remote sensing in the environment is a useful method in the management and protection of the environment. It gives off highly accurate data that will allow proper management of the environment.

8.15. DATA SCIENCE IN ENVIRONMENTAL MANAGEMENT

8.15.1. Environmental Data Collection and Management

Environmental Data Collection and Management System is a system using geographic information system and standardized data management procedures for site assessment appraisals, remediation actions and monitoring. It often uses multi-dimensional information analysis and data conception to support decision making processes for proper environmental mitigation procedures and remediation success.

There are many essential aspects to the gathering of environmental data as it relates to study design and sampling technique that will affect the form and strength of statistical inferences which are made: data collections helps with identifying the desired range of inference, choosing proper observational/experimental units, selecting the kinds of data to gather, and creating a robust sampling structure (that is., spatial, and/or temporal units distribution and method (s) of gathering the data) to ensure precise and precise inferences.

8.15.2. Big Data and the IoT Machine Learning

Internet of Things (IoT) is a combination of embedded technologies comprising of wired and wireless communications, actuator, sensor devices, and physical objects which are connected to the Internet. One of the long-lasting objectives of computing is simplifying and enriching human activities and experiences, or "computing for enhanced human experience."

IoT requires environmental data to either represent improved services to users or improve the IoT framework performance to achieve this logically. In this way, systems must be able to access raw environmental data from diverse resources over the network and evaluate this information in order to relevant extract knowledge.

Since IoT is among the most substantial sources of new data, data science offers a considerable contribution to ensuring IoT applications are more intelligent. Environmental data science is a mixture of different scientific fields which use data mining, machine learning, and other methods to find patterns and fresh insights from data. The techniques include a wide range of algorithms appropriate in different domains. The procedure of applying data analytics techniques to particular areas involves describing data types like volume, variety, and velocity; including data models like neural networks, classification, and grouping methods, and applying effective algorithms that equal with the data characteristics.

Because environmental data is produced from different sources with unique data types, it's essential to adopt or create algorithms that can manage the data characteristics. Secondly, the great number of resources which produce data in real-time aren't without the issue of scale and velocity. Ultimately, finding the best data model which fits the data is among the most important factors for pattern recognition and improved analysis of IoT data.

These matters have opened a massive number of opportunities in intensifying new developments. Big data is described as high-volume, high-velocity, and high assortment data that requires cost-effective, innovative types of information processing that allow enhanced insight, decision-making, and process automation. With regard to the challenges presented by big data, it's necessary to introduce a fresh concept termed smart data, meaning realizing productivity, efficiency, and value gains by using semantics to change raw data into smart.

A more recent description of this model is; Smart environmental Data provides value from harnessing the trials posed by volume, velocity, veracity, and variety of Big Data, while in turn providing practical information and enhancing decision making. Ultimately, smart environmental data can function as a great representative for IoT data.

8.15.3. AI to Improve Efficiencies

AI can help safeguard large forest zones by monitoring them through AI-allowed drones. The data gathered can be applied to map the zone and monitor changes. Including species detection, forest coverage zone, and poaching roads can be tracked through AI-enabled systems.

Using AI and IoT for environmental sustainability may help maximize current efforts towards environmental protection. Based on a 2018 study by Intel, 74% of 200 corporate decision-makers in environmental sustainability approved that AI would help resolve environmental problems.

8.15.4. Internet-of-Things

IoT offers operators real-time, data-based information concerning the building. Once you understand more about external impacts and human behavior, it's possible to make more precise temperature adjustments and improve your energy efficiency. This technology is well-proven, dependable, and constantly progressing to remain at the forefront.

IoT comfortably replaces traditional external temperature sensors. It obtains climate and temperature information, including advanced energy measurements for the particular property apart from connecting the building to AI's engine. This technology sends data to the sub-station which acts as a foundation for continual control, therefore allowing for optimum energy use including a perfect external climate.

There are also wireless sensors located across the apartments and within the sub-station. These sensors measure and dispense information on temperature and moisture levels in every apartment, including drift and return-temperature within the heating structure in the sub-station. This is something they do severally every hour, throughout the day. In case sensors or control structures from other energy providers have already been installed within the property, it means they can be applied to additionally optimize energy usage and indoor climate through connecting to data system. The unit is compatible with third-party sensors readily available on the marketplace.

Another creative feature is the IoT Ecosystem. It comprises of a renowned IoT platform which is connected to a cloud-based artificial intelligence (AI) software. Every data recovered and loaded through the firmware is processed through the app's self-learning algorithms, alongside other external information. This optimizes energy usage pro-actively and continuously. This technology is advanced though simple to apply and utilize. The connection just takes a few hours and is attuned with many other heating units and sensors found on the market. Connecting your unit with the environmental data forecasting ecosystem considerably increases the lifespan of your existent heating system. Being a property owner, you can save 10 to 20% of your overall heating costs with this product, plus the residents enjoy a better interior climate and the environment eliminates unwanted CO_2 gases.

8.15.6. AI Technology

With AI or advanced cloud solutions, environmental management technologies ensure the most ideal energy efficiency practices, whereas modern encryption functions and safe protocols ensure all data is well protected. The self-learning, cloud software can process more than 6 million data stations every day, with billions of metrics also from varying buildings in the database for constant analysis.

A given total sum of data that's continually increasing. The essence of the software largely is an algorithm which is continually refined and restructured. The more information the algorithm can analyze, the better it learns how a particular property behaves plus is influenced by different internal and external elements. The software hence provides intelligence for every day energy management, while the system's precision constantly improves as well.

8.15.7. IoT and AI for Curbing E-Waste

Millions of digital devices are castoff without proper disposal. Dollars are lost every year for proper discarding or recycling of used components of discarded devices. To alleviate the issue of inappropriate disposal of redundant electronic devices, businesses like Apple use recycled products or materials with a low harmful effect on the environment. The combination of hardware and software may prove useful for environmental protection.

Companies can use both AI and IoT for environmental sustainability alongside eco-friendly hardware to mitigate the production of e-waste. With appropriate techniques in place, the discarded devices may be refurbished and reused, therefore saving millions of dollars for the enterprise in creation of new products. Through IoT and AI, businesses can create systems to monitor their hardware. These systems can detect and notify consumers if the lifecycle of a part is complete, or has become faulty and lead the consumer through replacement processes. An effective IoT based AI-driven structure can help lessen carbon footprint.

8.15.8. Example of Artificial Intelligence in Environmental Management

eGain forecasting refers to a technique of managing building heating by measuring demand for heating-energy, which must be delivered to the building for every time unit. This is done by merging the dynamics of physical structures with meteorology, building properties, weather circumstances, outdoor temperatures, wind power and bearing, plus solar radiation.

As for conventional heating control, it's just the existing outdoor temperature that's considered; the beginning point for building the process of eGain forecasting is based on the ENLOSS arithmetic energy balance system established by Professor Roger Taesler, working for the Swedish Meteorological and Hydrological Institute in collaboration with Stefan Berglund and ThorbjörnGeiser from eGain Sweden AB.

Originally, the forecasting technique was introduced for use during the late 80s; but it wasn't until 2017 that it became inclusive. Currently, forecasting procedure has been applied in almost 16-million square meters of floor space of home buildings and commercial properties.

Since forecasting technique contains data about future demand plus isn't in any form of conflict with different approaches of improving energy efficiency, it's really a great foreground solution. In as far as applied use of forecasting technique is concerned, remote control forecast receivers normally are applied to dispatch and receive data through way of GPRS and GSM networks; besides the forecasting receivers supervise the systems of control panels fitted in buildings which regulate heat energy distribution, in the heating structure of a particular property.

Furthermore, special remote control climate loggers have begun to be used together with forecast receivers; the climate loggers measure air heat and moisture content with high accuracy plus the measurements are dispatched in real-time to forecasting receivers whereby they are connected.

This turning focus in technology means extra detailed preview of situations and increased potentials of control for operators of forecasting systems.

8.16. USING IOT AND AI FOR AGRICULTURAL PRESERVATION

Soil pollution is a key aspect faced today because of population growth, exhaustive farming, and other activities. Around 250,000 deaths happen every year throughout the world because of pesticide poisoning. Food generation is the goal of sustenance for human life on the earth. Application of IoT and AI for environmental sustainability comprises of methods used to monitor crops and soil, while maximizing crop production with a reduced impact on the ecosystem.

IoT and AI innovations have the capacity to transform conventional agricultural practices. They pave the path for safer agricultural approaches and ultimately benefit individuals' health. Smart monitoring equipment and sensors may be attached to crops to assess their growth constantly. Anomalies may be detected and resolved instantly. Parameters like hydration, plant nutrition, and illnesses can be examined in real-time. The data may then be utilized to decide irrigation patterns and recommend the most ideal watering cycles. A better range of crops having greater nutritional content can be farmed through AI, which can further lower usage of harmful pesticides.

8.17. ENVIRONMENTAL DATA ANALYTICS AND VISUALIZATION

At the core of science is the capacity to make logic out of data from observations or exhibited outputs. The challenge today is handling large capacities of data from various sources and to be able to detect insights in a simple to access way. Data visualizations play a key role in addressing this problem and helping with networking science outputs to a wide range of audiences. Well-designed data visualizations can certainly make a huge impact in being capable of uncovering discoveries that would otherwise stay unnoticed.

Interactive data science environments furthermore provide a way to quickly explore data, develop workflows, share codes, and visualize results. A range of Data Science environments are accessible, Jupyter is a common choice in the scientific world. It offers a well-supported open-source cross-language system for data examination and visualization.

Jupyter can further be extended by using standard configurations of IPython Widgets and even more advanced approaches to create customized Jupyter environments. Some examples of these works include: modified widgets for Pandas Data Frame Filtration, Customized widgets for environmental data and APIs Interactive Maps among others.

8.18. REMOTE SENSING AND GIS (OR SPATIAL DATA SCIENCE, MACHINE LEARNING, AND AI IN ADDRESSING ENVIRONMENTAL CHALLENGES) (VOPHAM ET AL., 2018) AVAILABLE AT: HTTPS://EHJOURNAL.BIOMEDCENTRAL.COM/ARTICLES/10.1186/S12940-018-0386-X

Geospatial artificial intelligence or (geoAI) is an emergent scientific discipline combining technologies in spatial science, AI techniques in machine learning (example, deep learning), data mining, and high-scale computing to mine knowledge from spatial big data. For environmental epidemiology, exposure molding is a commonly applied approach to perform exposure assessment to decide the distribution of exposures in research populations.

geoAI innovations provide significant benefits for exposure molding in environmental epidemiology, including the capacity to incorporate large quantities of big spatial as well as temporal data in various formats; computational efficiency; algorithms flexibility and workflows to accommodate appropriate characteristics of spatial (environmental) procedures including spatial nonstationary; plus scalability to model other environmental exposures across multiple geographic areas.

Spatial researchers apply technologies like geographic information systems (GIS), as well as remote sensing to spatial (example, georeferenced) data to realize these objectives-to detect and make sense of space patterns. Tied to the present era of big data is real-time production of spatial big data, which have gotten ubiquitously accessible from geotagged social media postings on Twitter to environmental sensors gathering meteorological information.

It's been suggested that roughly 80% of all data available are geographic in nature, since the majority of information everywhere can be georeferenced. With this measure, 80% of the 2.5 exabytes of big data produced every day is geographic. Data science, plus by extension spatial data field, are still developing fields that offer methods for organizing how we think around and approach creating new knowledge directly from (spatial) big data as it relates to the environment.

8.19. BENEFITS OF REMOTE SENSING AT HOME

8.19.1. Positive Weather Impact: Great Indoor Climate

In maintenance and renovation of buildings, it's important to check at every available opportunity for lowering heating costs. It would further be best if there was some way of compensating heat for drastic weather changes, which influences the property's elements to a big extent.

Remote sensing technologies like AI work brilliantly in homes since the installation process is simple, plus the investment is considerably low compared to other energy systems. Following installation, users will immediately notice an improved indoor climate, plus upon checking reports from Energy Experts, it's been found that savings perfectly match users' expectations.

That it's incredibly simple to save as much in energy is encouraging. With remote sensing, it feels great that people at the same instance can make positive impacts for the environment, and climate matters by decreasing CO_2 emissions.

As district heating prices continue rising going into the future, so will measures towards reducing energy consumption rates which shall probably increase heading forward. It's highly recommended to install home technology for this purpose. It's a small measure but with significant impacts on climate conditions, both within the house and outdoors.

8.19.2. Improved Heating Standards and Comfort

For better control of heating systems in a property, this technology would make an excellent installation. Energy is considered the largest expense factor for many property owners, thus it's greatly prioritized to find suitable alternatives which can lessen this cost so that it can become sustainable over the long run.

But the only thing which matters more is that residents must have the greatest comfortable indoor climates. The goal is certainly to find a good way of combining these particular priorities. Results show that the system can reduce energy consumption rate by up to 14.7 kWh/m^2, while also achieving a more balanced and comfortable interior climate.

8.19.3. Balanced Indoor Climate: Considerable CO_2 Reduction

For communities with conscious environmental mindsets, where people actively seek unique solutions which contribute to improved environmental preservation, remote sensing home technology may prove to be just the bet. After a swift installation, some of the results you'll enjoy include; reduced heating costs of around $23, 000 per year, lesser CO_2 emissions by up to 24,216 kg a year, reduced kWh/m^2 per year of 27 and energy savings of 19% a year.

The building will enjoy instant energy savings after data forecasting installations, and a significant advantage with it that you enjoy a more level indoor climate that's essential for members. Additionally, environmental data forecasting philosophy and management aims to "pair" and adjust heat supply to imminent external conditions resourcefully, therefore it doesn't become as susceptible to rapid weather shifts since the heating method is pro-active and rests in advance, plus the residents enjoy an even and enjoyable indoor climate, irrespective of the climatic conditions.

8.19.4. Great Visualization

It comes with inbuilt 3D models that can be used across the homes, these are great tools for visualization an area including how indoor climates at different venues, times, and conditions are reading. This is quite helpful when you want to follow up, or if you need to show how a particular situation is to other people. Visualization provides a one-stop solution for managing and getting feedback about the climatic conditions of your indoors and adjacent surroundings.

8.19.5 Transparent Heating. Cost Effective Housing

The system continually communicates temperature readings of every respective heating circuit, plus the rate of energy applied to the data forecasting consumer portal, which consequently visualizes the information. The energy specialist will continuously examine the collected gathered, this ensures that the heating structure's functional quality is excellent.

In comparison to traditional work schemes, which require janitors to make systematic visual examinations of the heating apparatus and bi-annual conservation visits through heating contractors, the technique creates a nearly real-time sequence of reporting on practical variations of heating systems.

With remote forecasting, it's possible to discover actuator motor failures within the heating structure of a segment of your property. When this constant monitoring is lacking, it's impossible to notice these motor faults until much later. Through the system, it's possible to identify and repair such malfunctions in the heating unit quickly before they worsen. This will prevent greater unnecessary damage to the system. Furthermore, in preventing this malfunction, they also save considerable energy costs. Energy management systems mechanically use weather forecasting for connected locations, plus has resulted in yearly energy savings of around 11.8% up to date. It's a great amount which helps the environmental technology to continue providing reasonable housing for people into the future."

8.19.6. Energy Management

Developers are constantly taking active approaches towards energy efficiency, including comfortable living conditions, construction technology solutions plus a balanced heating structure having a substantial impact on these elements.

The appropriate solutions and controls generally means that costs would be kept in balance, without ever compromising on comfort. Furthermore, comfortable living conditions are especially significant in senior housing facilities. Remote sensing, in turn, provides exceptional tools for monitoring and assessing heating usage in the property.

Among the key benefits of energy management is that room-temperature sensors make it practical for monitoring situations on a continual basis. Environmental data technology further reveals hidden data and validates previously acquired data, concerning the role of a building's internal heating system.

Also provided are practical and clear historical data for particular periods. Residents' usually report concerns with regards to living conditions, and are regularly dealt with skillfully and thoroughly, besides the number of such reports have been falling because of more level temperatures. Users also appreciate the fact they can speak easily to energy experts at any moment.

CHAPTER 9

Drones and Environment

CONTENTS

9.1. INTRODUCTION

The recent advancements in technologies are being utilized in the management of the environment. It being utilized in the protection of the environment. Drones are machines that have the ability to fly. They are small aircraft that do not have pilots in them. They are usually controlled by an individual on the ground. Some of the drones have cameras and are able to produce images of the areas of interest. In some instances, there are sensors attached on the drone. This allows them to pick up information about certain parts of the environment. It may pick up data on certain variables such the environmental humidity, temperature, and air pressure among others. With this information, researchers are able to know information about certain things such as the level of pollution in the land.

Drones provide a fresh and new innovative way of managing environmental issues. There are a variety of activities that can be conducted through the use of drones. It also allows the accessibility of areas that have been considered to be inaccessible areas and also areas deemed to be dangerous. Drones are among the latest trends in environmental management. When drones were introduced to be used in environmental management, there were certain concerns. Some of them included if drones would impact the environment in a negative way. It was later realized that the positives of using drones are more than the downside. There are a variety of ways which drones can be used in matters environment.

Some of the ways include:

9.2. ENVIRONMENTAL MONITORING

One of the ways in which individuals are able to get information concerning the environment is through monitoring of the environment. Drones are one of the ways in which environmental monitoring has been done. It has facilitated access to areas that have been termed to be inaccessible. A good example is when a researcher wants to know the undergrowth in a rainforest. The presence of thick undergrowth makes it difficult for one to gain information. In this situation, drones will be quite useful as they will be able to access and navigate through the thick undergrowth. The drones will facilitate the mapping of the forest. This will be useful in conservation purposes. It will also aid in the identification of illegal logging in forests that have been marked as protected forests.

The use of drone technology is also useful in remote areas. This applies to both onshore and offshore areas. Initially, remote sensing technologies were used in those areas. This involved the use of satellites as well as aircrafts to obtain information. The information obtained was not as through when placed in comparison with drones. The drones provide data that is more up close. This is made possible through the use of footage which are utilized in precision mapping. Drones have also been used as part of remote sensing techniques. This is attained by the attachment of sensors and thermal imaging equipment. The drones then become the platform. The drones are then used in various landscapes such as forests, waters, and ocean.

Environmental monitoring requires the analysis of collected data to gauge the environment. To conduct this, drones may be utilized as data collecting tools. This implies that will have the ability to collect data such as weather as well as soil. The data obtained will be used in effective land management. This allows the owners of land to get a better view of the environment which is their forests. This will be useful in the optimization of planting. It will also be useful in the monitoring the health of the vegetation. Drones have been useful in environmental monitoring through the use of aerial photography. They have the ability to take picture with high quality resolution. They are therefore useful in nature monitoring.

They have the ability to take crystal clear images of the various areas of the environment. Most of the drones used in environmental monitoring are fitted with meteorological tools including wind gauges, pressure sensors and thermometers. They are therefore able to collect data on climate change.

A major concern in matters environment has been the glaciers. The issue of climate change has greatly affected the glaciers where by researchers have seen much of the ice melting. The rate at which the glaciers are turning to water is alarming. To monitor the rate at which the glaciers are melting is through the use of drones. They are able to get images of the glaciers. By so doing, researchers are able to do a comparison of the glaciers at a certain age to a given duration before. By so doing they are able to monitor the rate at which the glaciers have melted. With this information, they are also able to know how much the rate of global warming has increased and the measures that should be employed to protect the environment.

In some of the countries, there has been recent occurrences of fires. This fire have been affecting large tracks of land. The satellites are not able to get a comprehensive information. Drones can be utilized in the monitoring of the areas. The information may be given to fire fighters who will be able to control the fire. When studies are conducted about the fires, the drones will be able to produce a photographic image of the area. The sensors will also facilitate the mapping of the areas that have been affected.

9.3. USE OF DRONES IN MAPPING OF THE CORAL REEFS

Recent studies have shown that most of the coral reefs have been affected by pollution among other things. This has generated the need for scientists to map out the coral reefs so that they can identify the areas that have been affected by pollution. They can be able to look for ways in which they can assist in the generation of the coral reefs. Since drones are affordable (both civilians and commercial users can access it) the scientists are able to utilize these tools. It allows them to get a better understanding of the coral reefs and they can be able to get more information on the ecosystems.

They are able to get more information since the drones facilitate the conductance of reef population surveys. The information allows the scientist to know the changes in the environment over a given period of time. They can also be able to identify the manner in which the coral reefs react to climate change. They can also know how pollution is affecting the coral reefs. The researcher will be able to measure the state and the structure of the difference. This is an upgrade from using technologies such as underwater photography. These methods were costly and also the images produced from these tools do not have a higher quality resolution.

This means that mapping of the coral reefs may be quite difficult. The drones are better as they are more detailed.

9.4. USE OF DRONES IN REPLANTING FORESTS

There is a great need for the replanting of forests. Increasing rates of global warming have called for drastic measures to be employed. The use of drones in matters environment has been used in the replanting of forests. Apart from researchers using drones in mapping of coral they have looked into the potential of drones being used in replanting of trees. This has been taken up by environmentalists. For then, this method is also cheaper and they are able to cover a larger ground. The method is also effective compared to other methods as per the research that has been conducted.

According to environmentalists, one drone has the ability over 100,000 trees. This is used by most of the reforestation companies. This can be achieved in one day. This means that if a number of drones are employed then the area covered will be much greater. There are drone seed companies that have looked into approved swarms for spraying. The company makes use of the drones to plant seeds. They also deliver water and also spray both herbicides and fertilizers. This method is advantageous for them as they cover large ground within a short duration of time. The company has also looked into using drones in the replenishing of mangroves.

This looks into using drones in planting the mangroves therefore causing a reforestation. This implies that there is the replanting of mangroves. This is a major area of focus because the mangroves are utilized by most areas to reduce impacts of various things including protection from hurricanes. It is also a good method of protecting the fish. The process involves the company using drones to map the area. They then determine the quality of the soil. They then identify the best areas that are suitable for planting trees as well as they species of trees that should be planted.

In the course of planting trees, there are drones that are specialized. They are specialized in the dispersing of seeds and they are fitted with instructions so that they ensure that the seeds have been fitted into the ground so that they make root.

9.5. USE OF DRONES IN COMBATING POACHING

Poaching has been a major issue of concern for environmentalists. This is because quite a number of animals are heading to extinction and this is attributed to poaching. The measures have been more so used in Africa where there are large numbers of animals that have been harmed through poaching. Most of them have become endangered. The environmental conservationist involved with animals have made used of drones in combating poaching. The major goal is to deter poachers. In Africa, the poachers poach the animals to obtain things such as ivory which is obtained from rhinos and elephants. Though most of the countries have been advised to make poaching illegal, some of them have not. This means that poachers get a market for the poached objects.

In most cases they are referred to as a big business as individuals are able to get lots of money therefore putting the animals in greater danger. As the conservationists cannot control those states, they have taken up measures that prevent poaching at the ground level. Some organizations have donated money to the conservationists that are utilizing the use of drones. They conduct both analysis and thwarting of poaching behavior. The recent data reveals that the use of drone technology has made an impact but not to a relevant extent. This implies that the yielded results are not promising as compared to the repopulation of trees as well as the coral reefs mapping.

The problem encountered when using drones to combat poaching is the identification and the catching of poachers. In most of African countries, there is a lack of a digital system that has a record of individuals in the poaching world. Though the drones will be able to identify the presence of an individual, it may not be able to identify whether the individual is a poacher or not. The drones are fitted with thermal imaging cameras or in some cases, they use infrared technology. This means that the drones can be utilized during the night which is the time where poachers are expected to kill the animals.

However, with the addition of this technology, it makes the drone heavier meaning that it may not be able to reach higher grounds. The other limitation of using drones is that they b require extended battery life and there is a challenge encountered when the drones are to be used in harsh weather or rough terrains. This contributes to the drones being more expensive. This is because when they crash, one will be required to purchase another drone making the equipment to be more expensive. It is essential that the drone pilot to have collaboration with local park rangers in efforts made to try and catch poachers when spotted by the drone.

The response by the park rangers will have to be fast once they are given information by the drone pilot. The rangers will therefore need to have experience and will require the needed tools in catching the poachers. Though the use of drones has been effective for some of the African countries due to the cooperation present from the governments, there are countries such as Namibia and Kenya which have banned to use of drones. Even for organizations who may take up the initiative to use drones in combating poachers, they will be required to have a license before they conduct any activities.

9.6. DRONES IN BIODEGRADABILITY

A major part of the environment is biodegradability. There are several cases where people have had their things ordered but delivered to the wrong person especially through the use of drones. The fear is that the drone crashing would have cause adverse effect to the environment. This is the reason why people have raised issues on the use of drones and its impact on the environment. To solve this problem, scientists have looked in to having the creation of drones that are biodegradable. There are some that have been actualized which include drones made of cardboard and drones made of microbes.

This is a good way to ensure the protection of the environment. If the parts are biodegradable, it means that when the drone happens to crash then it will be able to breakdown and assist in the improvement of the environment. This also reduced the number of plastics on the phase of the earth. Aerial Platform Support has been greatly involved in the creation of biodegradable drones. These are mostly made of carbon and are a one-time use drone. They have made use of DARPA's Vanishing Programmable Resource program. The companies intend their drones to be used in the delivery of supplies.

They will also be utilized in the delivering of medicines more so in remote areas and also in scenarios where there is a disaster that has occurred and there is great need for medicinal supplies. The drones are released from a ship or a plane and are dispatched to the area of need. Once they are released the drones are biodegradable within a given period of time. There are more innovations being made into making the biodegradable drones with a higher biodegradability. Youth in the tech world are also encouraged to come up with new biodegradable materials and create new generations of drones. They have to ensure that the level of technology used in the drone is maintained.

The recent trend in biodegradable drones has involved scientists in NASA creating drones through the use of fungus. The fungus used in this case is mycelium to be precise. It also utilizes chassis while the body of the drone is made of wasp's spit. The challenge has been to make both the motors and propellers to be biodegradable as they are not. More studies are being conducted to assist in the generation of the new generation of drones. There are other materials present in the environment that can be used in the making of these drones. It is only through conducting studies will the materials be identified.

9.7. USE OF DRONES IN SURVEILLANCE OF WIND TURBINES AND OIL PIPELINES

It is also used in monitoring of solar panels. One of the major ways of helping the environment is choosing to use energy from the environment. For instance, instead of using generated energy, one can opt to use solar energy. Most of the natural sources of energy need great monitoring so that the energy can be effectively harnessed and utilized. They also need to be inspected at a given frequency. The inspection is done to ensure that they do not break done incurring major costs. This has been the major reason why people have been reluctant to take up the use of the natural sources. This is attributed to the fact that it will require much of manpower and time also. This means that people have utilized mined minerals to produce energy which would have caused a major damage to the environment.

With the recent trend of the use of drones, it has facilitated the use of drones in the conducting of the inspection and monitoring of the natural sources of the environment. The drones will reduce the manpower. With the introduction of drones, people have taken up the use the natural energy sources. This has impacted the environment in a positive way and has minimized the rate of pollution of the environment. It has also benefited the environment in that it has facilitated the minimization of the use of heavy truck which would lead to damaging of roads.

In matters pipeline, the use of drones allows the coverage of a wide area of land within a short time. It will allow the identification of pipes that have been broken therefore reducing the chances of a pipe leakage. This will protect the various ecosystems from oil leaks. This will also contribute to the saving of resources.

9.8. USE OF DRONES IN WILDLIFE CONSERVATION

Wildlife conservation is really important in the preservation of natural spaces and also organisms. Environmentalists and researchers are working together to look for ways in which they can conserve the environment. They have looked into using of drones in the conservation of wildlife and to some extent it has proven to be effective. The drones have been used in the observation of large swaths of territory. This implies that the impact will be minimized on the area as few number of people will trek the area. The drones give off an aerial bird's-eye view. This allows the viewing of the animals while at the ground. The drones are usually fitted with cameras which allow the production of images.

They also allow the tracking of animals more so the dangerous animals. This minimizes the risk that would have been place if a human would have conducted the tracking. The drones are used to conduct aerial counts of the animals allowing ecologists to do a count of animals and keep record. They are able to identify which animal in endangered and those that need special care. The use of drones to some extent has been used in improving security.

9.9. USE OF DRONES IN THE TRANSPORTATION AND DELIVERY INDUSTRY

In matters, most of the pollution has been said to come from exhaust from cars. This type of pollution mostly affects the air aspect of the environment. The pollute on in the environment can also be attributed to transport in that most of the oil spills that have been incurred have been linked to damage while the oil was on transit in tankers or the oil pipes bursting due to heavy tankers passing over them and damaging the pipes. Some of the online delivery companies have decided to use drones in the delivery of their goods. This allows the delivery of these products at the doors of their customers ensuring that the product reaches the intended person.

Studies have shown that the environmental costs of pollution are relatively high and they have an impact in a variety of areas. Some of these areas include the air, highway, and railway. In the case of air transportation, it has been said the environmental cost is the smallest in areas of demand. The impact of air travel in the environment includes the emission of various gases nitrogen oxides. These oxide has a significant impact on the environment as they are released very high from the ground in the higher atmosphere. Though drones cannot effectively affect the use of aircrafts in terms of transportation of heavy cargo or human beings, it has made a significant impact in the delivery of products within short distances.

This has included drones being used in the delivery of pizza among other products. This will result to a drop in the air pollution which would have been caused by emissions from truck and cars. This is because they will be replaced by battery powered drones. This is among the contributions that drones have brought in the reduction of air pollution. Drones are, however, limited to the delivery of products of small loads. There are advancements being made to technology so that drones will be able to conduct the transportation of much heavier loads over relatively larger distances. This will result in a greater impact on the environment as batter powered machines will be used in the transportation of the loads.

In the case where the batteries used in drones are powered by green energy, the amount of carbon that is emitted from transportation tools will be reduced. The reduction will lead to the join in the fight against climate change. In the case of pollution of water and soil due to oil pipes bursting, this will also be minimized. If the drones will be used in the transportation of heavy cargo, then there will be no need for heavy trucks on roads meaning that the probability of the pipes bursting will be minimized.

9.10. USE OF DRONES IN LAND MANAGEMENT

Till recently, the major challenge encountered by most individuals is being able to have a better perspective on matters land. This is because things such as air surveys are not easily conducted. When it comes to building, the land is altered to accommodate the building and one has to ensure that the finished product is good enough. One cannot do this on the basis of imagination. They will be required to have an aerial view of the area. This is made possible through the use of drones. Conduction aerial surveys has been said to be a delicate art. Drones have been fitted with cameras to allow them take photographic images. Most of the drones have the ability to produce images of high resolution.

Through the use of drone technology, there is easy identification of the problem at hand. Drones have the ability to create 3D maps of the given areas. They also record the coordinates of the places therefore facilitating easy tracking of the area. It also reduces the man power that will be needed in the conducting of surveys as much of the work is conducted from the air. Through the latest trends in technology, most of the drones have the ability to do so many tasks as they are fitted with cameras and software that guides the entire process.

9.11. DRONES AND ROBOTS IN ENVIRONMENT

This involves the use of robots and drones in the protection of the environment. By using these two latest trends in technology, so much can be attained. This will cut across various fields. Ever since scientists and environmentalist have looked into the perspective of using robots in assisting in the protection of the environment, then why not consider the use of both the robots and drones in doing so. There are robots that have been specially designed to conduct activities such as cleaning up of oceans among others. The special modification is done so that they can be able to work in those areas without malfunctioning.

The use of both robots and drones open up new opportunities more so in research. For instance, they can be used in the collection of images of the environment while under water or above water. There are projects underway into the production of robotic vehicles which have tracking abilities and will minimize incidences such as spillage from occurring. There is also the option of using drones (aerial) as well as the robotic vehicles in facilitating a greater change in the fight against pollution. There is the generation of unmanned autonomous vehicles which will be utilized in environmental monitoring. Once the various states and governments pick up the use of drones to assist in environmental protection, it is estimated that there will be a significant change in the environment.

With the government's support, there will be funding of projects aimed at using drones to allow the protection of the environment. It will also encourage the private organizations involved in environmental protection to look into the potential of them using drones so that greater heights are attained. There is an estimation that the use of drones and robots in environmental monitoring is projected to grow as more initiatives are taken up in the assessing and minimization of pollution.

9.12. USE OF DRONES IN ENVIRONMENTAL RESEARCH

Research is an important area that should be looked into when there are efforts being put in place to assist something. In the case of the environment, there is plenty of research being conducted. For instance, in the rejuvenation of coral reefs, there is plenty of research that is done before any measures or steps are taken. In the case of conducting research, it is important that the researchers get data on whatever they are conducting research on. Take the case of a remote areas such as the sea rock where there are plenty of tourists who come to the area.

For researchers more so biologists, the problems have been to track the animal population in that given area. The initial methods employed when counting the population of certain animals would include the researchers rowing out and confirming the inhabitants of the sea rock. This method was not entirely accurate as there some species that would be recounted and so the information obtained would be false to some extent. The other challenge was that it was causing some distraction to the lounging locals. With the introduction of the use of drone technology, the situation has changed.

Researchers are able to use the drones to perform activities such as headcounts and also capturing images of the shaggy sunbathers. This method has proven to be much easier and effective in the long run. For instance, they are able to perform counting of fur seals with the use of a drone which has proven to be easier and effective in comparison to using boats and going to the sites. Environmentalists are able to do headcounts of the seal rock and also the Australian sea birds. They are able to keep track of the population of these animals and conduct analysis among other activities.

For scientists, the footage obtained from doing an aerial survey is said to give better insight on issues such as population change over a certain period of time. They are able to conduct comparisons noting if there was a decline or an improvement in the population of the animals. They have also been able to same on time as with the use of drones they are able to cover large tracks of land within a short period of time. The use of drones has also caused a reduction in the disturbance that would have been caused by the locals. The drones allow more space to be given to the locals and it does not disrupt them as they conduct their daily activities.

Once the researchers obtain data from the drones such as images or thermal data, the researchers process the information on the data obtained. They then conduct an analysis based on the obtained data. They are afterward able to get results on the analysis that was conducted. With this information, they are able to identify endangered areas. They can also come up with strategies or measures that they can incorporate and use in combating the problems encountered in the environment. This way they assist in improving the environment.

For the biologists and ecologists, they are finding the use of drones to be more helpful and also, they are very good assistants. In line with research, there technological advancements being made on drones so that they are able to assist in clarifying manners in which climate change is affecting the earth. They are also using it in combating ecological problems. Some of these measures have included the development and incorporation of an onboard software that will allow the detection of green fishing netting seals as they usually get wrapped up in. By so doing they are able to get information on seals in real time. They can then make appropriate decisions concerning the seals and the appropriate response they should take.

In the case where the net causes the seals to get tangled up, the researchers will be able to know that, take up a boat and look into ways in which they can free the seals. The ability of drones to access remote areas is very useful to scientists. Researchers are at a better hand of understanding the changes in weather patterns in areas such as the rainforests, mountain area and underwater ecosystems. Some of the researchers have used drones in measuring the surface reflectivity. They are able to log in how much solar energy a particular landscape reflects and absorbs. Some of them have taken up the use of UAV's in the surveying of vine climbing plants. This enables them to get a better understanding on how they affect the carbon balance in the tropical rainforests. They are able to take record of the carbon dioxide levels in the atmosphere in certain periods.

In the case of the underwater ecosystems, the researchers have been utilizing drones by sending them to collect sediments samples ad also do an analysis on the climate change's impact on ocean life. There is the use of wind-powered Sail drone fleet which conducts patrols in the high seas. They then produce data which is relayed in real-time. A sail drone unit has the ability to operate indefinitely. This factor is attributed to its design and sonar capabilities. It also has the ability to harness renewable energy. The fleet is modified so that it has the ability to measure radiation, chlorophyll, pH levels, and salinity among others. Under the name of research, the researchers under the sailing drone are looking forward to sending the drones into hurricanes. This will allow them to have a better understanding of the behavior and the impact of the storms in different areas.

The ground-based drones are utilized in assisting in minimizing deforestation, plant seeds, and coastal erosion. The drone seed works in post-fire areas conducting replanting of trees in the given areas. There are also drones that have been utilized in drone planting and they work in that they have the ability to launch seed pods from 10 feet above the ground.

Bibliography

1. Barrett, E. C., (2013). *Introduction to Environmental Remote Sensing* (p. 480). London: Routledge.
2. Bharagava, R. N., (2020). *Emerging Eco-Friendly Green Technologies for Wastewater Treatment* (pp. 1–77). Berlin: Springer Nature.
3. Chang, N. B., (2012). *Environmental Remote Sensing and Systems Analysis* (pp. 35–85). Florida: CRC Press.
4. Cheremisinoff, N. P., (2002). *Handbook of Water and Wastewater Treatment Technologies* (pp. 26–101). London: Butterworth-Heinemann.
5. Craven, J., (2019). *High-Tech Solutions for Flood Control: How Engineers Mitigate Rising Floodwaters*. Retrieved from thoughtco.com: https://www.thoughtco.com/how-engineers-stop-floods-177699 (accessed on 13 July 2020).
6. Denboba, M. A., (2005). *Forest Conversion, Soil Degradation, Farmers' Perception Nexus: Implications for Sustainable Land Use in the Southwest of Ethiopia* (pp. 1–89). Göttingen, Germany: Cuvillier Verlag.
7. FAO, (2000). *Manual on Integrated Soil Management and Conservation Practices* (p. 214). London: Food & Agriculture Org.
8. Fernquest, J., (2011). *Flood Prevention Technology*. Retrieved from bangkokpost: https://www.bangkokpost.com/learning/learning-news/268638/flood-prevention-technology (accessed on 13 July 2020).
9. Ghosh, S. N., (2016). *Environmental Hydrology and Hydraulics: Eco-technological Practices for Sustainable Development* (p. 426). Florida: CRC Press.

10. Hellin, J., (2016). *Better Land Husbandry: From Soil Conservation to Holistic Land Management* (pp. 1–81). New Jersey: CRC Press.
11. Huber, J., (2004). *New Technologies and Environmental Innovation* (pp. 3–59). London: Edward Elgar Publishing.
12. Hussain, C. M., (2020). *The Handbook of Environmental Remediation: Classic and Modern Techniques* (p. 506). London: Royal Society of Chemistry.
13. IAEA, (2012). *Getting to the Core of Environmental Remediation: Reducing Radiation Exposure from Contaminated Areas to Protect People.* Vienna, Austria: Department of Nuclear Energy International Atomic Energy Agency Vienna International Centre.
14. Johnson, D. M., (2006). Feasibility of water purification technology in rural areas of developing countries. *Journal of Environmental Management*, 3–21.
15. Kalogirou, S. A., (2009). *Solar Energy Engineering: Processes and Systems* (pp. 1–121). New York: Academic Press.
16. Lafond, A., (2020). *The Ultimate Guide to Air Purifiers Technologies.* Retrieved from foobot: https://foobot.io/resources/the-ultimate-guide-to-air-purifiers-technologies/(accessed on 13 July 2020).
17. Maczulak, A. E., (2010). *Sustainability: Building Eco-friendly Communities* (p. 241). New York: Infobase Publishing.
18. Mati & Bancy, M., (2006). *Overview of Water and Soil Nutrient Management Under Smallholder Rain-Fed Agriculture in East Africa* (pp. 9–43). Colombo, Sri Lanka: IWMI.
19. Nelson, V., (2013). *Wind Energy: Renewable Energy and the Environment,* (2nd edn., pp. 1–65). Florida: CRC Press.
20. OECD, (2011). *OECD Studies on Environmental Innovation Invention and Transfer of Environmental Technologies* (p. 237). London: OECD Publishing.
21. Oppenländer, T., (2007). *Photochemical Purification of Water and Air: Advanced Oxidation Processes (AOPs)-Principles, Reaction Mechanisms, Reactor Concepts* (pp. 1–79). Florida: John Wiley & Sons.
22. Pichtel, J., (2019). *Fundamentals of Site Remediation* (pp. 1–79). Maryland: Rowman & Littlefield.

23. Pimentel, D., (2008). *Biofuels, Solar and Wind as Renewable Energy Systems: Benefits and Risks* (pp.1–79). Berlin: Springer Science & Business Media.
24. Sayigh, A. A., (2012). *Renewable Energy, Technology and the Environment* (p. 3500). Sydney: Newnes.
25. Singh, R., (2006). *Hybrid Membrane Systems for Water Purification: Technology, Systems Design and Operations* (pp. 1–57). New York: Elsevier.
26. Sioshansi, F. P., (2011). *Energy, Sustainability and the Environment: Technology, Incentives, Behavior* (pp. p. 640). New York: Elsevier.
27. Skidmore, A., (2017). *Environmental Modelling with GIS and Remote Sensing* (pp. 1–68). Florida: CRC Press.
28. Speight, J. G., (2019). *Natural Water Remediation: Chemistry and Technology* (pp. 3–91). Berlin: Elsevier.
29. Spellman, F. R., (2014). *Environmental Impacts of Renewable Energy* (p.478). Florida: CRC Press.
30. stormwater.wef.org, (2016, January 7). *New Technologies Help the Environment Agency Prevent, Prepare, and Respond to Flooding.* Retrieved from stormwater: https://stormwater.wef.org/2016/01/new-technologies-help-environment-agency-prevent-prepare-respond-flooding/(accessed on 13 July 2020).
31. Sutherland, K. S., (2004). *Profile of the International Filtration and Separation Industry: Market Prospects to 2009* (pp. 1–75). Netherlands: Elsevier.
32. Talwar, P., (2012). *Manual of Assisted Reproductive Technologies and Clinical Embryology* (pp. 15–78). New Delhi: JP Medical Ltd.
33. Weng, Q., (2011). *Advances in Environmental Remote Sensing: Sensors, Algorithms, and Applications* (pp. 3–57). Florida: CRC Press.
34. Xu, Z., (2013). *Fundamentals of Air Cleaning Technology and Its Application in Cleanrooms* (pp. 1–47). Berlin: Springer Science & Business Media.
35. Xu, Z., (2018). *Air Purifier: Property, Assessment and Applications* (p. 143). Berlin: Springer.
36. Yeoshua, S. B., (2005). *Environmentally Friendly Technologies for Agricultural Produce Quality* (p. 552). Florida: CRC Press.

INDEX

C

D

E

F

G

H

I

L